AF817794

# Quantitative Methodologies using Multi-Methods

# Quantitative Methodologies using Multi-Methods

## Models for Social Science and Information Technology Research

*Sergey V. Samoilenko and Kweku-Muata Osei-Bryson*

Routledge
Taylor & Francis Group
NEW YORK AND LONDON

First edition published 2022
by Routledge
6000 Broken Sound Parkway NW, Suite 300, Boca Raton, FL 33487-2742

and by Routledge
2 Park Square, Milton Park, Abingdon, Oxon, OX14 4RN

© 2022 Taylor & Francis Group, LLC

Routledge is an imprint of Taylor & Francis Group, LLC

Reasonable efforts have been made to publish reliable data and information, but the author and publisher cannot assume responsibility for the validity of all materials or the consequences of their use. The authors and publishers have attempted to trace the copyright holders of all material reproduced in this publication and apologize to copyright holders if permission to publish in this form has not been obtained. If any copyright material has not been acknowledged please write and let us know so we may rectify in any future reprint.

Except as permitted under U.S. Copyright Law, no part of this book may be reprinted, reproduced, transmitted, or utilized in any form by any electronic, mechanical, or other means, now known or hereafter invented, including photocopying, microfilming, and recording, or in any information storage or retrieval system, without written permission from the publishers.

For permission to photocopy or use material electronically from this work, access www.copyright.com or contact the Copyright Clearance Center, Inc. (CCC), 222 Rosewood Drive, Danvers, MA 01923, 978-750-8400. For works that are not available on CCC please contact mpkbookspermissions@tandf.co.uk

**Trademark Notice:** Product or corporate names may be trademarks or registered trademarks and are used only for identification and explanation without intent to infringe.

*Library of Congress Cataloging-in-Publication Data*
A catalog record for this title has been requested

ISBN: 978-0-367-90396-1 (hbk)
ISBN: 978-1-032-04697-6 (pbk)
ISBN: 978-1-003-02414-9 (ebk)

Typeset in Times
by KnowledgeWorks Global Ltd.

# Contents

# Preface: Possible Uses of this Book

It is only expected that our reader should be presented with a possible explanation, before actually reading this book, for why this text should be read at all. Sir Winston Churchill once quipped that he only read for pleasure or profit, hence, following this binary option exercised by this well-regarded individual, we suggest that it is a profit that our reader could derive from reading the material we present in this book.

Specifically, we suggest that our work could be used for five different, but not mutually exclusive, purposes.

First, our reader may approach the book as one of the means for a user-friendly introduction to the popular methods of data mining and data analysis. In our presentation and overviews of the methods, we avoided getting involved into details that are more suitable for more advanced users – we assume that our audience has, at most, surface-level knowledge of the methods presented in the book. The rationale was that once a reader gets an idea, a conceptual grasp, regarding how a method works, then she will be able to get a more detailed overview from other sources. After all, it is not an exaggeration that a separate book could be written (and they are written) in order to give justice to each of the methods presented in our text.

Second, given a reader's introductory knowledge of the methods covered in this text, our book could serve as an introductory guide to the subject of complementarity of the tools and techniques of data analysis – we aim to demonstrate how the methods could be used in synergy in order to offer insights into the issues that could not be dissected by any single method alone. We do not suggest that the methodological modules we describe in the book are the only combinations possible. Instead, we advise our reader to create her own modules, while keeping in mind a necessity of justifying the selection of methods and arguing the beneficial nature of their interaction to her own audience. After all, the resultant methodology has to make sense not only to the investigator, but also to the audience that is interested in the subject. On a side note: we found that the structure of the type "Assumption/Justification" works well for the purpose of making any assumptions and assertions explicit and for eliciting a rational explanation for why the selected methodology is one that is worthy of consideration.

Third, we suggest that this text could be used as a set of templates, where, given a set of research questions, the investigator could identify a set of methodological modules allowing for answering the research questions of interest. This is not entirely unlike the relationship between analysis and design phases of the systems development life cycle – where the "What?" of the analysis phase has to be translated into the "How" of the design phase. So, we hope that our reader, with the help of this book, will be able to identify methodological modules (the "How") that are suitable for answering her research questions (the "What"). Simply put, we hope that our work would help in transitioning a conceptual domain of the research questions into a scaffolding of the data analytic and data mining methods by which the research questions could be answered.

Also, this text could serve as a guide to exploring what the data holds. Let us say, a researcher is in possession of a set of data, complemented with domain area' knowledge. Usually, there are frameworks and theories that could be used to conduct scientific inquiries, but, sometimes, it is not clear what sort of questions the data may answer. In such case (which is somewhat similar to grounded research-type of an approach), the investigator may use the methodological modules presented in this book (corresponded with the running examples) to generate a set of preliminary questions which, after a careful consideration and a requisite culling, could be formulated into a set of questions consistent within a selected theory or a framework.

Finally, we suggest that our work could be used as a generator of new research questions. This is because, as a reader would have an opportunity to discover, the result of applying every method in each of the modules opens up a new dimension ripe with follow-up questions of the type "Why is this so?" Consequently, even if an inquiry started with a set of research questions based on a theory or a framework, at the later point the study could become augmented by the additional follow-up questions that arose as a result of the application of the selected methodology.

# Introduction

An investigator undertaking a research project usually faces a variety of challenges to overcome, the following four being the most obvious – we suggest them without implying that their order is an indicator of a relative significance. The first challenge is associated with the expression of the overall goal of the study, corresponded by the formulation of a set of well-defined research question(s). The second challenge is that of selecting the underlying framework of the study, where the level of granularity of the research should be supported by the appropriate level of precision, or detail, of the research framework (e.g., level of grand theory vs. level of a case study). The third challenge is the availability of the data. Finally, there is a challenge of selecting or designing an appropriate methodology that allows for analyzing the available data and answering the research questions of the study.

We have dealt with challenges of creating theoretical frameworks in our prior work (e.g., Samoilenko & Osei-Bryson, 2017), and now we concentrate our effort on the methodological aspect of data analytic inquiry. Commonly, it is a case that the decision regarding the appropriate methodology of the study, as well as the decision concerning the selection of the data analytic methods, is impacted by the nature of the available data. It is fair to say that the question of the quality and availability of the data is unique to each study. However, the general guidelines are straightforward – get as much of the relevant good quality data as possible. In some cases, the primary data must be collected by the investigators, while in other cases the secondary data could be obtained from reliable public sources.

In this text our aim is to illustrate how quantitative, multi-method methodologies could be applied to answer a variety of research question. As a result, our intent is to offer our audience a general map, in a form of a methodological framework, that could be applied to a variety of research scenarios with the purpose of crafting and answering research question that are fairly typical of quantitative inquiries based on the analysis of non-parametric data.

We intend to achieve our goal by demonstrating a creation of what we call methodological modules – synergistic combinations of two or more data analytic techniques that could be used independently, or combined together to create complex methodologies. We are aware that the investigators need to defend the validity of the methodology of the study to their audience and the reviewers, and in order to assist our reader in this regard we offer some published examples and their citations for each methodological module we describe.

Samoilenko, S. and Osei-Bryson, K.-M. (2017). Creating Theoretical Research Frameworks Using Multiple Methods: Insight from ICT4D Investigations. Auerbach Publications. ISBN-13: 978-1498779951.

# Introduction

# Section I

## Development of the Methodological Modules

# 1 Pre-Requisite General Questions

In crafting a multi-method methodology, a "slice of an apple" analogy is applicable – each method is akin to a knife slicing an apple in its own way under a different angle. The greater the variety of slices, the richer the perspective we can obtain on how an apple looks inside. The question to answer is, of course, how to arrange the process of slicing in a complementary way, especially if we are dealing with a basket of apples, and, furthermore, with a basket of apples of possibly different kinds. Let us consider our "basket of apples" analogy and consider some of the questions that could be asked.

*Do we assume in our investigation that the apples in the basket are of the same kind?*

The positive answer to this question reflects the assumption of *homogeneity* of the sample. Consequently, we are looking, conceptually, for a single apple that is representative of the whole basket. And, of course, we don't assume that our basket of apples is just like all other basket of apples out there in this world – we don't generalize our findings beyond the basket.

*Do we assume that the apples in the basket are of different kinds?*

The basic assumption stemming from a positive answer to this question is that of *heterogeneity* of the sample. Resultantly, the obvious question to pose will be regarding the number of different *kinds* of apples in our basket. This is something that needs to be discovered, for we don't have any "received" knowledge neither regarding the categories of apples in the basket, nor regarding the number of the categories.

*Do we know that the apples in our basket are of different kinds?*

The answer to this question comes in the form of the received knowledge expressed in the form of categories defining heterogeneous sub-sets of the apples in our basket. The important point to make here is that the set of categories is not necessarily definitive and final; instead, it is simply a categorization based on the value of a chosen by the external agency' attribute. For example, we can be told that the basket of apples consists of the sub-sets based on color. However, we may also be told that the sub-sets differ in terms of size. Or, in terms of taste (e.g., sweet, tart, and sour), or, in terms of anything else that can be usefully applied for the purposes of partitioning the sample (e.g., perceived attractiveness and freshness).

## IMPACT OF THE ASSUMPTION OF HOMOGENEITY OF THE SAMPLE ON RESEARCH QUESTIONS

The questions above could be considered to be a "first order" questions, and after answering them we could proceed to generating much more interesting, "higher order" questions. Clearly, the most limited in terms of the follow-up questions is the option based on the assumption of homogeneity of the sample. Fundamentally, our "higher order" questions are fairly simple, and are of descriptive-predictive nature. For example, we could, assuming the homogeneity of the sample, ask the following question:

*What is a common set of attributes and their values describing the basket of apples?*

Additionally, given a time series data, we could pose a question such as:

*What are some of the changes that took place over time that impacted the apples in the basket?*

This, fundamentally, would allow us to pose a pretty much the last interesting question of the scenario, namely:

*Given the changes that impacted our basket of apples over time, what are some of the changes that will impact our apples in the future?*

Given the scenario described above, it is easy to see that answering such question does not require a multi-method methodology; instead, a single method will do just fine.

Unlike the assumption of homogeneity, however, the assumption of heterogeneity of the sample gives rise to a much richer set of "higher order" questions.

For example, in the absence of the received knowledge regarding the categories, an investigator may ask the following question:

*What are some of the sub-groups that are present in the sample?*

A "basket of apples" scenario may offer an answer of the type: The basket of apples consists of three groups of apples – red, yellow, and green.

The very logical follow-up to such questions is the question regarding the nature of the differences of the sub-groups, namely:

*Given the presence of multiple sub-groups in the sample, what are some of the factors responsible for the heterogeneity?*

Keeping in mind our example, we can come up with such answer, as: an apple variety is responsible for the membership in a group. Or, the levels of chlorophyll and carotenoids impact the apple' group membership.

Similarly, in the presence of the received knowledge regarding the differentiation of the sub-groups, a researcher may ask a question of the type:

*Given the sub-sets of the sample based on the received criterion, what other factors/attributes are associated with differentiating criterion?*

Again, in addition to the difference in the level of pigments that determine the group membership, an investigator may discover that the levels of exposure to sun and of the temperature also do impact the color of apples.

This may lead to an inquiry into the factors that differentiate the sub-groups the most, as well as the least.

As we can see the assumption of heterogeneity of the sample gives rise to a set of the related research questions, where each question will require an application of its own data analytic method to answer.

At this point we can state a very simple rule of thumb: *an assumption of heterogeneity of the sample gives rise to the application of multi-method methodologies.*

## FROM A BASKET OF APPLES TO A SET OF SYSTEMS (DECISION-MAKING UNITS)

A much more sophisticated set of "second order" questions could be generated if we consider our sample to be comprised not of a basket of apples, but of a basket of *systems*. For our intents and purposes, we define a *system* as a *structurally and functionally complex entity that receives inputs and transforms them into outputs.* And, of course, systems that are of interest to us are open, dynamic, non-linear, and complex. For all intents and purposes economies, departments, hospitals, universities, NBA players, assembly line workers are all examples of systems that could be represented/referred to as units that transform inputs into outputs.

This gives us another three dimensions to consider, namely, level of inputs, level of outputs, and efficiency of conversion of inputs into outputs. It goes without saying that "input" and "output" are logical designations – a system could have multiple inputs and multiple outputs. Under the assumption of homogeneity of the sample we can generate, for all intents and purposes, three types of "second order" questions, for example:

*What is the average (or, min, max, etc.) level of inputs?*
*What is the average (or, min, max, 25th percentile, and so on) level of outputs?*

Additionally, this basic model describing the system from the perspective of inputs, outputs, and the process of transformation opens up an access to the questions associated with the *relative performance* of the system. This leads to the questions of the type of:

*What is the level of performance of the system – the level of average (smallest, greatest) level of transformative capacity (conversion of inputs into outputs)?*

It is easy to see, in the case of the assumption of homogeneity of the sample, that such questions could be answered via using a very simple methodology – a single descriptive statistics-type method will do the job.

The methodological simplicity disappears once we operate under the assumption of the heterogeneity of the sample. In addition to the mentioned above three questions, which will need to be answered *per sub-group*, we can generate such additional questions as:

*What are the differences in inputs/outputs/transformative capacity between groups?*
*What are the sources of the differences?*
*Is the difference in outputs is due to the difference in inputs, or is it due to the difference in efficiency of the process of transformation?*
*What are the reasons for the differences in transformative capacity?*

If the investigator is in the possession of the time series data, then another set of questions related to the *performance of the system over time* could be generated, such as:

*What are the changes in the level of performance of the system over time?*

Once the changes have been identified, an investigator may be interested to find out the answer to the follow up question, namely

*What are some of the reasons for changes in performance of the system over time?*

Intuitively, it is easy to see that answering "second order" questions will require employing a methodology that is not based on one or two data analytic methods. However, the complexity of a suitable methodology increases even more if we consider systems in their contexts.

## FROM SYSTEMS TO SYSTEMS IN CONTEXT

The consideration of a context of the systems – *decision-making units* – gives rise to even more sophisticated set of, what could be called "third order", questions. Such inquiries, fundamentally, deal with investigating the impact of the environment on the relative level of performance of the system and could be generalized in the form of a following question:

*What is the impact of the environmental factors on the level of the relative performance of the system?*

The basic premise behind such question is that the environment impacts the performance of a set of heterogeneous systems differently. And, of course, this leads to a significant increase in the number of attributes describing the system, for now we need to consider not only inputs, outputs, and transformative capacity but also the factors that could plausibly influence them from outside.

A simple illustration is provided by a comparison of various types of engines – any and all engines require some sort of a fuel, produce some form of energy that is applied to do work, and require a transformative capacity to convert fuel into energy. It is easy to see that temperature, humidity, air pressure, as well as other factors, have a capacity to impact fuel, the process of conversion of fuel into energy, as well as relative amount of work produced by the generated energy. Resultantly, an investigator could ask the following "third order" questions:

*What environmental factors influence inputs of the system?*
*What environmental factors influence outputs of the system?*
*What environmental factors influence the transformative capacity of the system?*

The above mentioned questions and the scenario applies to the case when we consider the environment of our sample of units to be homogeneous. We can ask even more interesting questions if operate under the assumption that our systems reside in different environments – when we are dealing with a heterogeneous sample comprised of sub-groups, where each subgroup resides in its own specific context.

Let us consider a sample consisting of retail stores – the previous scenario would allow us to compare the stores located in approximately the same climate and operating within similar socio-economic areas/neighborhoods. However, what if we decide to compare the levels of relative performance of an athletic shoe store located in California with a work boots store operating in Alaska? Given this scenario, an investigator will need to find out the environmental factors common to both stores. It is important to note that by "environmental factors", we denote a wide variety of those factors that are *external* to the common "input-output" model describing the stores. Consequently, a researcher investigating the factors influencing the difference in the performance of the systems/units functioning in different contexts may ask a following question:

*Do heterogeneous contexts of the sub-groups in the sample share common factors?*

Once this question is answered, the follow-up question begs to be asked, namely:

*Do common environmental factors have a similar impact on the level of the relative performance of the systems functioning in heterogeneous contexts?*

It is easy to see, by now, that an inquiry addressing such question would have to rely on a methodology comprised of multiple methods of data mining and data analysis.

# 2 Components of Multi-Method Methodologies

At this point, we would like to provide our readers with a very brief and informal overview of the data analytic tools that will be used to create what we call "methodological modules" – logically related collections of methods that could be applied in order to answer a variety of complex research questions. The following context is not intended to be a thorough introduction, but, rather, a jargon-free and novice-friendly introduction that should be followed by a more comprehensive coverage we offered in our previous work.

## CLUSTER ANALYSIS (CA)

The purpose of CA is to test the data set for the presence of heterogeneous subgroups. So, if we have a sample of National Hockey League (NHL) players, then CA may offer us insights regarding the similarity of the hockey players in our sample. Of course, it is always possible to assume that the data set consists of homogeneous entities, that is, if one uses criteria allowing for such assumption. For example, a set of NHL players could be considered a homogeneous based on the fact that all of the members of the set are, indeed, members of NHL teams, or, they are all hockey players, or, because they all have been drafted and given a draft number.

But, most of the time it is useful to search for characteristics of hockey players based on more granular differentiation. For example, the NHL players may differ in regard to the position they play, based on the years of experience, based on height and weight, or any multitude of factors that could be of interest to the investigator. However, the problem with creating sub-sets of the sample is based on the choice of criteria, for, in the case of the self-selected criterion, a researcher would almost always face a reviewer's question: Why did you select this criterion?

CA, being an unguided technique, helps in identifying sub-sets of the sample in a fashion that is somewhat independent of the researcher. An "unguided technique" means that CA partitions the sample based on the criteria of the method itself – CA will attempt to identify the presence of the clusters in such way that the members of each cluster are similar to each other, but the clusters are different from each other. As a result, we are getting groups that are "similar inside and different outside". It is always possible, of course, that CA will not produce any clusters – this will happen if the data set is comprised of truly homogeneous entities.

The "truly" part, however, carries a connotation of subjectivity – who determines what "truly" means? After all, it is easy to make an argument that the differences are there, but an investigator simply assumed their absence. And this is where an investigator influences the results of CA – via the selection of settings according to which clustering will be performed. Let us consider the most popular types of CA.

*K-means* clustering approaches the data set with the requirement to identify $K$-sub-groups within the sample, where $K$ is chosen by the investigator number. For example, if we have 100 NHL players in our sample, then the selection of $K = 3$ will result in three sub-groups, the selection of $K = 2$ gives us two, and of $K = 4$ will give us four sub-groups of hockey players.

*Divisive hierarchical clustering* approach is based, unsurprisingly, on dividing, in a top-down fashion, a complete set into a number of clusters, where the stopping criterion is set by an investigator. So, the starting point of this approach is a complete data set that is being partitioned in a step-by-step approach until a specified criterion is met.

*Agglomerative hierarchical clustering*, on the other hand, starts with the consideration of every member of the set to be, so to speak, its own cluster, and proceeds in a bottom-up fashion, to combine – to agglomerate – the members of the set into a larger set of clusters. Both hierarchical approaches operate in a step-by-step fashion, and the investigator is given an opportunity to see the intermediate results of each step – one can actually see the process by which the clusters are formed.

Regardless of the choice of the clustering method, the result of CA is a set of sub-groups – clusters – comprising the sample. So, it is easy to see that while the investigator has an influence on the number of clusters, the basis for clustering is method-driven and unguided – the selection of the variable based on which CA is performed is not under the control of the investigator. Meaning, if the results of CA yielded three clusters, then we know that the sample is comprised of three groups, but we don't know, really, what factor or factors are responsible for the heterogeneity of the sample.

An important point to consider is the relative size of clusters. Let us say, our sample of 100 NHL players was clustered into three groups. What should be a minimum size of the smallest group? Would the memberships such as Cluster 1 = 80, Cluster 2 = 17, and Cluster 3 = 3 be useful in the investigation? This is not an easy question to answer. Clearly, Cluster 1 and Cluster 2 are useful to consider, but Cluster 3 looks like an outlier. In our own research and our previous work, we suggested a rule of thumb for the smallest cluster to be at least 10% of the sample.

So, the results of CA yield a set of clusters comprising the data set and allow for answering the question of relative homogeneity of the sample. But, the results of CA do not allow for answering the question of what is the underlying reason for the differences. The next method, Classification Trees, allows for gaining an insight into the nature of the difference between the clusters.

## CLASSIFICATION DECISION TREES INDUCTION (CDTI)

There are two general reasons for why an investigator may want to use decision tree induction. One of the reasons is *to predict* the result of something – of a value measured on interval scale. For example, given a previously constructed model of the relationships between inputs and outputs of an NHL player, we would be able to predict a productivity (e.g., output) of the incoming player based on the initial conditions (e.g., minutes played per game, experience in the league, etc.). The second reason is *to classify* an entity according to the pre-defined categories. In this overview, we concentrate on a decision tree's analysis with the purposes of classification – CDTI. For example, given a general model of an NHL player, based on the categorization of "star", "mid-level", and "flop", we will be able to place a given player within one of the pre-defined categories. But the most important use of CDTI is to uncover the reasons for why the categories differ from each other.

Let us consider, again, a sample of NHL players that was partitioned into the "star", "mid-level", and "flop" categories. It is important to note that the reason for the partitioning is not important – this could be due to the received knowledge, or this could be due to the results of our own inquiry (e.g., cluster analysis). For example, we could also create groupings based on the position of the players – "defense" vs. "forwards" vs. "goaltenders", or based on the geographic area of the player's origin – "North America" vs. "Eurasia". Or, the sub-sets could be created based on such criteria as "number of year in the league" or "number of games played", or "time spent on injured list" and so on.

In the case of our example, grouping the players into "star", "mid-level", and "flop" categories gives us an opportunity to place a player within one of the three categories. However, the reason for why the categorization was made is not clear – we, really, don't know what separates "mid-level" from "flop", we simply know that a player is in one group, or in another. But what we really would like to find out is the difference between the groups – after all, the difference could be bridgeable in a given case, and if we could help transitioning "flops" to "mid-level", so much the better.

By creating a *target* variable, let us say, "Group", and assigning one value per group – "1" for "star", "2" for "mid-level", and "3" for "flop" – we can include this variable in our data set and run CDTI. So, if we had 20 variables describing NHL players, we will end up with $20 + 1 = 21$ variables. Once we designated the "Group" variable as the "target" variable, CDTI will attempt to construct – to "grow" – a decision tree. The decision tree will take the form of an upside-down model of a real tree, with the roots and a trunk representing a complete data set (e.g., 21 variables multiplied by whatever number of NHL players we got in the sample) that is being, in a step-by-step fashion, partitioned into branches and then leaves.

Once we generated the decision tree, there are two things of immediate interest (again, we remind our reader that in this overview we concentrate on the basics only), namely, what are top-level splits and what are the memberships of the top-level nodes. The top-level splits are important because CDTI "grows" the trees on the basis of the variables differentiating the nodes at each level the most. Thus, the top splits will be based on the variables that are the most important in splitting the data set into the groups. In the case of our NHL example, the top-level splits could be based on *scoring points* (a sum of goals and assists) and *time on ice* (a total time on ice in the current season). The knowledge of the two most important variables, as well as of the values, levels, of the variables producing the split helps us to identify the most important criteria differentiating "star" from "mid-level" from "flop". It is also helpful to consider that the generated decision tree could be presented not only in its graphical format but also in the format of *English rules*, which is a textual description of the tree. However, the practical and interpretive significance of the top-level split variables is impacted by the nodes' membership.

The relative membership of the top-level nodes – the sub-groups of the sample produced by the top-level splits – tells us how "clean" the separation between groups is. Meaning, we would like to get a series of splits placing our groups of hockey players neatly into their assigned categories. And, we can see that if we use *scoring points* and *time on ice* as the top-level split variables, then they should produce very clean separation between the groups based on the levels of the respective values. For example, we would reasonably expect that "star" will have high values in terms of *scoring points* and *time on ice*, "flop" will have low values, and "mid-level" will reside somewhere in between.

However, let us imagine that we don't have in our possession a set of variables that are designed to differentiate the groups – we don't have stats to separate NHL players into the performance-based categories. Consider utilizing such attributes as height, weight, years of experience, and age. Then, we could easily see the results of CDTI being much less clear – we will see, if the top-level splits are made based on height and weight, a much more mixed picture, where each node of the decision tree will be populated by the mixture of "star", "mid-level", and "flop" representatives.

## NEURAL NETWORKS (NNs)

A NN is another technique that is often used for the purposes of prediction, but, unlike decision trees induction (DTI), a NN allows for predicting values of multiple variables. There are two basic requirements for using a NN: First, the data set must be comprised of interval data, and, second, the data set must be relatively large. Structurally, a NN could be described as a system comprising three parts: A set of *input nodes*, a set of *output nodes*, and a *black box* in between. The contents of the black box are comprised of the collection of one or more layers of *inner*, or *hidden*, *nodes*.

The purpose of input nodes is to provide a set of values that are considered to be not necessarily input values to a process of transformation, but, rather, represent a set of starting conditions. For example, in the case of NHL players, we could specify such inputs as height, weight, age, years of experience as inputs. Here, we can see that such attribute as height is not an input to any process of transformation, but it is simply a one of the "givens" – something that a decision maker could designate as one of the points of departure.

The purpose of output nodes is to specify, similarly, not the result of transformation of inputs via the transformative capacity of the system, but the end result of something of interest. In our example of hockey players, output nodes could be represented by the minutes played, goals, assists, and so on. Clearly, one would be hard pressed to explain, for example, the process that produces such value as "minutes played" for a given hockey player.

The sole purpose of the hidden nodes is to connect, via a "black box" model, a set of input nodes to the set of output nodes. Structurally, hidden nodes are represented by weights that are derived/calculated, or obtained, during the process of "training" of a NN. Functionally, hidden nodes serve the purpose of translating inputs into outputs – not shedding light on the process of transformation, not explaining, not revealing anything regarding the mechanism of transformation, but simply supplying a collection of weights allowing for converting inputs into outputs.

Let us consider an example of hockey players with the input nodes comprising {height, weight, age, years of experience in NHL} and the output nodes being {minutes played, goals, assists}. Then, for a member of our data set, the actual *Input* → *Output* model would look something like "(6.2, 195, 29, 8) → (22, 0.8, 1.1)". So, given our set of input nodes and output nodes, the hidden nodes of our NN would have to "figure out" how to transform left side into the right side. And this "figuring out" process is the process of *training* of NN, during which the weights of the hidden nodes are calculated.

This process of training a NN is done on the training sub-set of the sample, and once a NN is constructed, it would be *validated* on the testing sub-set to check the applicability of the model to a larger set. So, after we trained and tested a NN, we get the model of the type "inputs → transformation → outputs", where, again, "transformation" component is meaningful to the model, but meaningless to an investigator. However, once it was generated, it could be used for two (and here we concentrate on the most obvious ones) interesting purposes.

First, we could predict the outputs of new entities that were not a part of the original set. If we consider a new NHL player, then we could use the developed NN model to predict, based on the original conditions, his performance in terms of the predefined outputs. We would not be able to say how the results (outputs) are obtained, or what the player should do to improve the outputs, but we will be able to come up with the outputs based on the NN model's "black box" of hidden nodes.

The second scenario involves building two NN models and requires a little introduction. Let us consider two groups of NHL players – "flops" and "stars". "Flops" play less, and "flops" score and assist less. "Stars" play more, and "stars" score and assist more. If we are to consider the situation of "flops", then there are two plausible explanations for their inferior performance. First, they score and assist less because they play less. So, if we want to improve their performance, we have to let them play more – as much time as the "stars" get to play. Then we'll have a level playing field for a comparison. But, second, it is also possible that "flops" score and assist less because they are simply inferior players. And before giving "flops" more playing time, they have to improve their level of skills. What is the proper way to go about improving "flops" performance?

If we have two NNs generated – one for "stars" and another for "flops", then we could use it to gain sufficient insight to select a better way. First, we can save "black box" model of "stars" and apply it to the inputs of "flops". This is analogous to giving "flops" skills of "stars", and this will allow us in determining whether the poor performance of "flops" is due to their lack of skills. Second, we can apply the inputs of "stars" to the "black box" model of "flops". This is analogous to giving "flops" inputs of "stars" (e.g., giving more playing time), and this will allow us in finding out if the inferior performance of "flops" is due to their relatively smaller playing time. By comparing the values of the output nodes for both scenarios, we can determine the reason for the difference in the outputs and design an appropriate intervention accordingly.

## ASSOCIATION RULES MINING (ARM)

There is an interesting similarity between the ARM technique and NNs. Conceptually, both NN and ARM work with, or are comprised of, three fundamental components – left-side variables, right-side variables, and a "black box" in between. Unlike NN, however, which aims to connect *specified by the investigator* inputs of the left side with outputs of the right side, the goal of ARM is to discover possible associations between left-side variables and right-side variables that are *not specified a priori*. Thus, while the purpose of the "black box" of a NN is to *force* the connection between the pre-defined inputs and outputs, the purpose of the "black box" of ARM is to *discover* associations between the left-side variables and the right-side variables that are not selected in advance. In both cases, the "black box" offers no explanation regarding the nature of the relationship between the left and the right side, leaving this task under the purview of the investigator.

Our aim here is in applying ARM for the purposes of *Market Basket Analysis* (*MBA*), which gives us an opportunity to discover a set of interesting "If → Then" rules. We must note that "→" does not connote a causal relationship between the left side and the right side, but simply indicate that "If" and "Then" go together. For the purposes of MBA, we have to present our data set as a collection of transactions, where each transaction consists of a set of *items*. Commonly, a data set for MBA comes from the transaction-level data associated with purchases – "market baskets". Consequently, a data set for MBA is comprised of the records that are "baskets" of purchases, and we'll have as many baskets as there are records in the data set.

In simple terms, the purpose of MBA is to find out what items in the customer's basket tend to go together. Clearly, some items, while not necessarily are random purchases (e.g., those that are a result of spur of the moment decision), are caused to be purchased by random events. For example, a person shopping in a grocery store may get a pack of light bulbs (because prior to going to the store, he accidentally tipped over a table lamp and the light bulb broke) or a jug of a windshield washing liquid (because the warning sign came up in her/his car but the local auto shop is already closed). Such purchases are not very interesting for decision makers, because it is hard to predict them, as it is hard to predict random events.

However, some purchases do tend to go together – milk and bananas, salad mix and salad dressing, bread and cheese, as well as many other combinations of products. Clearly, in no way or form a purchase of one item causes, deterministically, the purchase of other items, but, nevertheless, they seem to go together naturally. Discovery of such rules is the main goal of MBA, for a decision maker is provided with the actionable information that is based on non-random events, and, therefore, is somewhat predictable.

ARM and MBA do not work with numeric data; however, a data set comprising the numeric values could be converted into a nominal data set that could be subjected to the analysis. We will demonstrate to our reader, later in the text, one of the ways to perform such conversion. One of the most popular algorithms for generating "if → then" rules (commonly referred to as *itemset*) is *A priori*, and commonly used practice of generating itemsets is based on the threshold values of three criteria – *support*, *confidence*, and *lift*.

*Support* of an itemset is an indicator of a frequency of occurrence of a given rule within the data set. If a value of support equals to 1, then the itemset appears in 100% of the records (transactions) in our data set. Clearly, a value of support could not be equal to 0, but higher the value within the range between 0 and 1, the greater the percentage of the records that contain the rule. A commonly chosen threshold value for support is 0.2, which means that the algorithm will generate itemsets that appear in at least 20% of the transactions (records) in the data set.

Once an itemset has been discovered by the algorithm, it needs to be tested – after all, the rule may, or may not be true. The test of how often the rule has been shown to be true within the data set is reflected by the criterion of *Confidence*. In the case of the value of confidence being 1, this means that all (100%) the transactions (records) containing "If" part also contain "Then" component. Similarly, if the confidence is equal to 0.1, then only 10% of the transaction containing "If" also contain "Then".

Support and *Confidence* are useful measures for evaluating a likelihood of occurrences of "If" and "Then" within the data set, but they are not good indicators in regard to "→". This is where *Lift* comes to play, being a criterion estimating a probability that "If" leads to, or, serves as an antecedent to "Then". If the value of *Lift* is equal to 1, then it means that "If" and "Then" are independent of each other; if the value is greater than 1, then it means that the presence of "If" results, actually, in the presence of less of "Then" and vice versa. Consequently, if the value of lift is less than or equal to 1, then the "If → Then" rule is not very useful to a decision maker. However, if the value of *Lift* is greater than 1 (the upper range is infinity), then the rule is important to the decision maker because, for all intents and purposes, there is evidence that the presence of "If" is positively associated with the presence of "Then". Simply put, the criterion of lift serves as an indicator of the importance of the discovered rule for the purposes of prediction.

Let us consider our running example of the set of NHL players and use of MBA to gain possible insights into the data. In order to generate association rules, we need to have a data set consisting of transactions; thus, we need to convert all the numeric stats into some sort of categories. For example, in regard to playing time, we could generate such categories as "low", "medium", and "high". Or, we could create categories based on four quartiles – "low", "mid-low", "mid-high", and "high". This process of finding a suitable categorization is under the purview of the investigator; thus, no "one size fits all" scheme exists.

Also, in order to generate a set of rules, we must have a much richer data set than the one of the type "minutes played → points". Instead, we need a data set where we would have a wide variety of attributes for each player. If we consider "minutes played → points" being an analytical model describing transformation of inputs into outputs for each player, then we should also consider that every player exists/functions in their own context – in their own environment. After all, the same "minutes played → points" model does not discern between offensive and defensive players. And the environment always plays an important role by impacting the level of performance of an open system. Consequently, we would like to have a data set that describes the context within which NHL players exist. This could be any data regarding the team where they play, the type of family environment they have, anything regarding their hobbies, pets, and preferences. We would definitely like to include any data describing the types of exercises they do, the types of diets they may have, and so on.

Once we have such a data set, we should be able to generate rules that may inform us, for example, about interesting associations that characterize "low" time hockey players. Similarly, we should be able to generate rules that associate some of the common contextual variables for "star"-level players. As a result, we could obtain some insights into contextual commonalities that the players of different types possess. Such rules, of course, would not allow for explaining how the "playing time" is converted into "points", but they will offer insights regarding what environmental variables are worth paying attention to.

## DATA ENVELOPMENT ANALYSIS (DEA)

DEA is a very popular method commonly employed for the purpose of calculating levels of relative efficiency of decision-making units (DMUs). As long as an entity could be described as a set of inputs and outputs, it qualifies as a DMU and could be subjected to DEA. The DMUs in the set do not have to be of the same size or scale, nor do they have to share a similar context. Instead, DMUs have to be *semantically* similar – they have to be of the same *type of systems* that receive inputs and produce outputs. For example, it would be reasonable to compare, from DEA perspective, different types of restaurants, airports, and factories, but it would be hard to defend a comparison of restaurants vis-à-vis airports vis-à-vis factories. Simply put, DMUs in the sample have to make sense to be grouped together for a meaningful comparison.

So, if we consider a group of semantically similar entities (e.g., hockey players, departments, firms, and hospitals) that are described by the same *DEA model* – by the same set of inputs and outputs, then we can subject that data set to DEA to compare their levels of relative efficiency. There are few points worth mentioning regarding a DEA model.

First, inputs and outputs of DEA are *non-monetary*, for it is not the purpose of DEA to accommodate a model of the type "investments → revenues". Instead, inputs and outputs are those other variables than costs and prices that make sense to the investigator. For example, we can compare efficiencies of a group of students using the following DEA model: "(#ofStudyDays, #ofHoursPerStudyDay) → TestGrade". Similarly, we would not use DEA to conduct an analysis of relative efficiencies of a group of bakeries using a model "cost of ingredients of a cake → price of a cake", but we could use DEA to investigate a model "(# of employees, # of shift hours) → # of pastries baked".

Second, the inputs and outputs are not necessarily representative of a production-type "Input → Output" model. For example, we would not be interested in a DEA model consisting of "weight of ingredients to make a cake → weight of a cake" to evaluate the level of a relative performance of a set of bakeries, but we may use a DEA model consisting of "(# of employees, # of hours of operation) → # of cakes sold". Once DEA modeling is adopted, all DMUs are considered to be completely described by the inputs and outputs of the model – no context-specific, environmentally relevant attributes are considered by the analysis.

Fundamentally, the approach of DEA is that of constructing a ratio – the multiple inputs and outputs of DEA model are collapsed/transformed into one meta-input and one meta-output, and the ratio of the two is expressed as a score of a relative efficiency. It is important to note that DEA does not compare each DMU with the rest of the DMUs in the sample, but only with a suitable sub-set comprised of its peers. The DMUs with the highest values for the ratio are considered to be relatively efficient as compared to relatively inefficient DMUs in the set. All relatively efficient DMUs receive a perfect score of "1", which indicates a 100% efficiency. Once a sub-set of relatively efficient DMUs is identified, they form an efficiency frontier that envelops the rest of the relatively inefficient DMUs.

Another aspect of DEA that is worth mentioning is that of orientation of the model. DEA allows for three orientations: *Input-Oriented, Output-Oriented*, and *Base-Oriented. Input-oriented* model is used in the scenarios where the inputs are controllable (e.g., students can control how many days a week and how many hours per day they study), *Output-Oriented* model is suitable for situations when the outputs are controllable (e.g., bakery can control how many cakes it is going to produce), and *Base-Oriented* model assumes the control over the inputs as well as the outputs (e.g., a writer has a control over how many hours a day she is going to dedicate to writing, as well as she can control the number of pages written). Regardless of orientation, all relatively efficient DMUs receive scores of 1, but relatively inefficient DMUs will receive scores of less than 1 in the case of input orientation, and greater than 1 in the case of output orientation.

Additionally, there are options regarding return-to-scale, where DEA offers to an investigator consideration for constant, variable, and non-increasing (decreasing) returns to scale. The choice of an option impacts the shape of the efficiency frontier, so, for example, a relatively efficient DMU under constant return to scale is also efficient under variable return to scale, but the opposite is not necessarily true.

DEA is a *point-in-time* method, where calculations of the scores of relative efficiencies of DMUs are undertaken for a chosen temporal snapshot. However, it is also possible to use DEA to investigate *changes* in the scores of relative efficiencies of the DMUs over time. This is done via calculating values of *Malmquist Index* (MI), which measures the changes in scores that took place over a period of time – let us say, year 1 and year 2. Fundamentally, values of MI are calculated by conducting DEA twice – once for year 1 and the second time for year 2. MI is a very useful tool, for it allows an investigator to assess whether, over a period of time, a given DMU became more efficient (positive change in productivity), stayed the same (no change), or became less efficient (drop in productivity). This, on its own, is very valuable, for we can assess (let us continue using our example of NHL players) not only relative efficiency of the player in regard to conversion of minutes played into points but also whether a hockey player became better, stayed the same, or became worse over a period of time.

There is a great additional benefit offered by MI to an investigator – it allows to identify the *sources of the change* that took place over time. Let us consider the approach of DEA – the approach of constructing an efficient frontier at a given point in time. For all intents and purposes, DEA tells us how far away from the efficient frontier a given DMU is situated. But the approach of MI is based on measuring the movement of the DMUs over time, and there are two components that comprise the movement.

First, there is a movement of the efficiency frontier itself – this movement is represented by *Technology Change* (TC) component of MI. Second, there is a movement of a DMU relative to the frontier – this is *Efficiency Change* (EC) component of MI.

If we, again, consider our example of NHL players, then MI would be able to provide us some very valuable insights. Let us say, that a given player exhibited a positive change in his score of relative efficiency over the period of one year – his value of MI is greater than 1. What caused the change? Well, it is possible that the positive change was a result of the movement of the frontier – the player was "dragged" up by the technological advancements that took place over time. The skates became better, the stick became lighter, a better helmet, and so on. But it is also possible that the positive change was a result of the improved skills of the player – he improved his skating, he improved his puck handling, and so on. By calculating MI and its components, we would be able to find the reason for the improvement. Similarly, if a given player decreased its score of the relative efficiency, then we'll be able to identify the sources responsible for the deterioration.

Clearly, MI is a useful tool used for the purposes of designing an intervention directed toward improvement of the level of relative efficiency of DMU, because it offers a justification for the targeted allocation of resources. For example, if a given hockey player exhibited a decrease in the level of the relative efficiency over time (e.g., $MI<1$) under the scenario of the positive change in technology (e.g., $TC>1$) and the negative change in efficiency (e.g., $EC<1$), then the appropriate intervention would be toward spending more time and money working on the skills of the player, and not toward buying the latest and greatest equipment.

As it was mentioned earlier, a DEA model does not assume a causal relationship between the inputs and outputs of the model. However, it may be worth testing. And in this situation, Multivariate Regression may become a very useful tool to apply.

## MULTIPLE REGRESSION (MR)

The purpose of the MR is to model a relationship between multiple independent variables and a single dependent variable – in our overview we leave out covering the concepts of the intercept and error term of the MR. Interestingly, it is possible to "build a bridge" between DEA and MR, albeit an imperfect one. We attempt to do so via finding some similarities between the two methods and pointing out some obvious differences.

The main difference between the two methods is, of course, an assumption of normality of the data set. MR assumes a normal, bell-curve, distribution of the data in the sample, and the sample is being representative of the general population. DEA, being a non-parametric tool, requires no such assumption. This does not mean that MR could not be performed on the non-parametric data, it simply means that the interpretation of the results will be limited to the sample and is non-generalizable beyond the data set at hand.

Another difference is in regard to the nature of the relationship between inputs and outputs. While DEA is a "black box" model that is not concerned with how, exactly, inputs are converted into outputs, MR assumes the causally linear relationship between inputs and outputs of the model. Additionally, while the purpose of

DEA is to inquire into the *relative efficiency* of conversion of inputs into outputs, the purpose of MR is to inquire into the *strength of the relationship* between inputs and an output.

Just like DEA, MR requires the presence of input variables, but unlike DEA, MR is limited to a single output (dependent) variable. This does not mean that MR cannot be applied to the DEA model, it simply means that in order to do so, the DEA model would be translated into as many MR models as there are output variables. For example, if we consider a DEA model "(A, B, C) → (D, E, F)", then we can translate it into the following three MR models: "(A, B, C) → (D)", "(A, B, C) → (E)", "(A, B, C) → (F)".

Just like MR model, a DEA model could be comprised of multiple input variables and a single output variable. But unlike a DEA model which, let us recall, collapses input variables into a single meta-input variable, MR model considers the input variables separately – this allows for assessing the individual impact of each input variable on output.

Generally, the MR procedure estimates a linear relationship within the following model:

$$Y = a + b1 * X1 + b2 * X2 + b3 * X3 + \ldots + bn * Xn + \varepsilon$$

where Y = dependent variable; a = intercept; b# = slope coefficient; X# = independent variable; ε = error term.

MR relates each independent variable in the partial fashion to a dependent variable. Meaning, a coefficient of each independent variable represents partial contribution to the dependent variable, while controlling for the rest of the independent variables in the MR equation. As a result, MR correlates the dependent and independent variables in a partial fashion, which is often referred to as a *partial correlation*.

Somewhat similar to the DEA model, the MR model considers input variables to "work together" to manifest their impact on outputs. In both cases, an investigator, implicitly or explicitly, makes an argument for why inputs do belong together. This brings the question: If they do belong together, then are there any complementarities between them? Unlike DEA, which does not allow answering such question, MR allows for including interaction terms into MR model that helps answering such question.

Let us consider the following DEA model "(# of employees, # of shift hours) → (# of pastries baked, # of orders filled)". For all intents and purposes, this model could be converted into two MR models: "(# of employees, # of shift hours) → (# of pastries baked)" and "(# of employees, # of shift hours) → (# of orders filled)". This, in turn, could be extended to include an interaction term allowing to test the presence of complementarity between the number of employees and hours worked:

1. "(# of employees, # of shift hours, *# of employees X # of shift hours*) → (# of pastries baked)"

and

2. "(# of employees, # of shift hours, *# of employees X # of shift hours*) → (# of orders filled)".

In the case of the first model, we'll get the following:

# of pastries baked = a + b1*# of employees+ b2*# of shift hours +b3**# of employees * # of shift hours* + ε.

The test for the significance of the impact of *# of employees* and *# of shift hours* on *# of pastries baked* requires us to test the following null hypotheses:

H01: b1= 0
H02: b2= 0.

And the test for interaction – the significance of the impact of *# of employees* and *# of shift hours* on *# of pastries baked* amounts to testing the null hypothesis

H03: b3 = 0.

The interpretation of the interaction term in MR, however, is not as straightforward as the interpretation of the slope coefficient (e.g., b1 and b2) of an independent variable. For example, b3 in the equation above reflects the relationship between *# of pastries baked* and *# of employees* and *# of shift hours* when *# of employees* and *# of shift hours* increase jointly. Furthermore, b3 in the equation above reflects conditional relationship between dependent and independent variables, for the impact of *# of employees* on *# of pastries baked* would depend on the level of *# of shift hours* and vice versa. In general, answering the question of the presence of complementarity would require the investigator to inquire into such potentially affecting the interaction term issues as the presence of thresholds, level-dependent dynamic of the interacting variables, and so on.

In order to show the benefits that MR could offer to an investigator, let us take a look, again, at our example of NHL players. Let us suppose that we are interested in some of the factors that impact playing time. And let us consider practice and training time, per week, being factors of interest. Then we can formulate the following MR model:

"# minutes of play time per game=*f* (#of minutes of weight training, # of minutes of ice practice, # of minutes of running)".

By running the MR, we will be able to find out not only whether independent variables on the right impact the dependent variable on the left, but we also will be able to find out whether there is a complementarity between the certain types of training. Clearly, we will not be able to generalize the finding beyond our sample (e.g., to the sample of National Collegiate Athletic Association [NCAA] hockey players, or high school hockey players, etc.), but this will not be a problem if the sample we got represents the population we are interested in.

# 3 Framework for Methodological Modules

Now that we briefly covered some of the more popular data analytic methods, we can start putting together a general guideline for using the methods in synergy. In Table 3.1, our reader can find a brief summary of the insights and limitations of each of the methods covered in this book. Table 3.2 is dedicated to two-method modules, while Table 3.3 outlines the same for three- and four-method modules.

Again, it is worth repeating that we consider each of the methods, as well as their combination, from the perspective of their application to systems. This leads to two scenarios: first, the case where the "input→output" model of the systems in the sample is considered explicitly (e.g., via specifying a DEA or NN model), and, second, where no model is considered. Figures 3.1 and 3.2 offer possible scenarios for each case.

At this point, we are ready to present to our reader appropriate to the scenario methodological modules. We will proceed in the order of increasing complexity of the modules, as follows:

A. Homogeneous sample
   1. DEA and Decision Trees
   2. DEA and Association Rules

B. Heterogeneous sample (groupings are given)
   1. Decision Trees and Association Rules
   2. Decision Trees and Multivariate Regression
   3. Decision Trees, DEA, and Association Rules
   4. Decision Trees, DEA, and NNs

C. Heterogeneous sample (groupings are not known)
   1. Clustering and Decision Trees
   2. Clustering and Association Rules
   3. Clustering, Decision Trees, and Multivariate Regression
   4. Clustering, Decision Trees, and Association Rules
   5. Clustering and DEA
   6. Clustering, DEA, and ARM
   7. Clustering, Decision Trees, and DEA
   8. Clustering, Decision Trees, DEA, and NNs

**TABLE 3.1**
**Insights and Limitations of Single Methods**

| Method | Offered Insight | Limitation |
|---|---|---|
| CA | Allows for testing an assumption of homogeneity of the sample and identifying presence of sub-groups in the sample. | In the presence of multiple sub-groups does not offer any insights into the sources of heterogeneity. |
| DTI | Given the target variable, allows for identifying attributes responsible for differentiating sub-groups of the sample. | Target variable must be provided "from outside". Does not consider impact of differentiating variables on an "input→output" model of sub-groups. |
| NN | Allows for creating "input→output" model of transformation. | A "black box" model of the input-output conversion process. Does not allow for evaluating the model in terms of the relative efficiency. Does not allow for assessing the impacts of individual inputs on outputs. |
| ARM | Allows for identifying a set of "if→then" rules present in the data set. | Does not provide any insights regarding an "input→output" process. |
| DEA | Allows for calculating scores of the relative efficiency of DMUs, as well as changes in the scores (via using MI) over time. | A "black box" model of the input-output conversion process. Does not offer insights into the sources of inefficiencies. |
| MR | Allows for determining the significance of the impact of independent variables on a dependent variable and identifying the presence of complementarities. | Does not provide any insights regarding an "input→output" process and does not allow for considering multiple outputs of the process. |

**TABLE 3.2**
**Insights and Limitations of Two-Method Combinations**

| Module | Offered Insight | Limitation |
|---|---|---|
| DEA and CA | Allows for identifying naturally occurring sub-groups in the sample, and analysis of the relative efficiency of the sub-groups. | Does not explain the nature of the differences between the sub-groups and offers no insights into the reasons for the differences in the scores of relative efficiency. |
| CA and DTI | Allows for identifying naturally occurring sub-groups in the sample, and for identifying the sources of heterogeneity. | Does not consider the presence of an "input→output" model. |
| DTI and ARM | Allows for classifying sub-groups based on the received criterion and for generating the "If→Then" rules. | Does not consider naturally occurring groupings of the sample and does not consider the presence of "Input→Output" model. Relevance of "If→Then" rules to the classification is not clear. |

| Module | Offered Insight | Limitation |
|---|---|---|
| DTI and MR | Allows for identifying naturally occurring sub-groups in the sample and for assessing the impact of multiple input variables on a single output variable. | Does not allow for discovering naturally occurring groups in the sample and for assessing complex "input→output" model comprising multiple outputs. |
| DEA and DTI | Offers insight into differences between relatively less and more efficient DMUs. | Insights are limited to a few variables (top-level splits), and offers no insights regarding the relationships between the identified variables. |
| DEA and ARM | Allows for generation of the causal structures unique to relatively efficient and inefficient groups. | Does not allow for identifying a subset of relevant causal structures based on the set of criteria that differentiate the groups. |
| DEA and MR | Allows for calculating the scores of the relative efficiency of DMUs, and for assessing a significance of individual inputs of the DEA model on individual outputs. | Does not offer any insights regarding the nature of the differences between more- and less-efficient DMUs. Does not allow for considering the context of DMUs. |

## TABLE 3.3
## Insights and Limitations of the Developed Methodological Modules

| Module | Offered Insight | Limitation |
|---|---|---|
| DEA, CA, and DTI | Allows for identifying the factors associated with the differences in the relative efficiencies of the sub-groups in the sample. | Does not offer any insights regarding the source of the differences, and yields no causal structures differentiating the groups. |
| CA, DTI, and MR | Allows for discovering naturally occurring sub-sets of the sample, as well as for identifying the sources of differences between sub-sets. Allows for assessing the differences in the impact of independent variables on a dependent one between sub-groups. | Does not allow for assessing context-specific differences between the sub-sets. Assumes linear relationship between independent and dependent variables. |
| CA, DTI, and ARM | Allows for discovering sub-sets within the sample and for identifying the sources of heterogeneity. Helps uncovering common to the sample, as well as group-specific itemsets. | Offers no opportunity to identify the relationships between the sources of heterogeneity and discovered itemsets; offers no insight into the structure of the itemsets. |

*(Continued)*

**TABLE 3.3 (Continued)**

| Module | Offered Insight | Limitation |
|---|---|---|
| CA, DEA, and ARM | Allows for identifying naturally occurring sub-sets of the sample and for calculating the scores of the relative efficiency of the members of each sub-set. Helps uncovering common to the sample, as well as group-specific itemsets. Allows for considering important to the DEA model context-specific factors. | Does not explain the sources of heterogeneity of the sample. Provides no insights into the relationship between DEA model and itemsets. |
| DTI, DEA, and ARM | Helps identifying the sources of heterogeneity of the sub-sets of the sample. Allows for calculating the scores of the relative efficiency of each sub-set. Helps uncovering common to the sample, as well as group-specific itemsets. | Sub-grouping of the sample is based on the received knowledge. Offers no insights into relationships between the sources of heterogeneity and DEA model, or group-specific itemsets. |
| DTI, DEA, and NN | Helps identifying the sources of heterogeneity of the sub-sets of the sample. Allows for calculating the scores of the relative efficiency of each sub-set. Allows for identifying the reasons for disparities in the scores of relative efficiency between sub-sets. | Sub-grouping of the sample is based on the received knowledge. Offers no insights into relationships between the sources of heterogeneity and DEA model. Utilizes two "black box" models offering no insights into the process of transformation. Does not consider interrelationship between individual inputs and outputs. Does not allow for considering the role of a context. |
| CA, DTI, DEA, and NN | Allows for discovering sub-sets within the sample and for identifying the sources of heterogeneity. Allows for calculating the scores of the relative efficiency of each sub-set. Allows for identifying the reasons for disparities in the scores of relative efficiency between sub-sets. Allows for considering important to the DEA model context-specific factors. | Offers no insights into relationships between the sources of heterogeneity and DEA model. Utilizes two "black box" models offering no insights into the process of transformation. Does not consider interrelationship between individual inputs and outputs. Does not allow for considering the role of a context. |

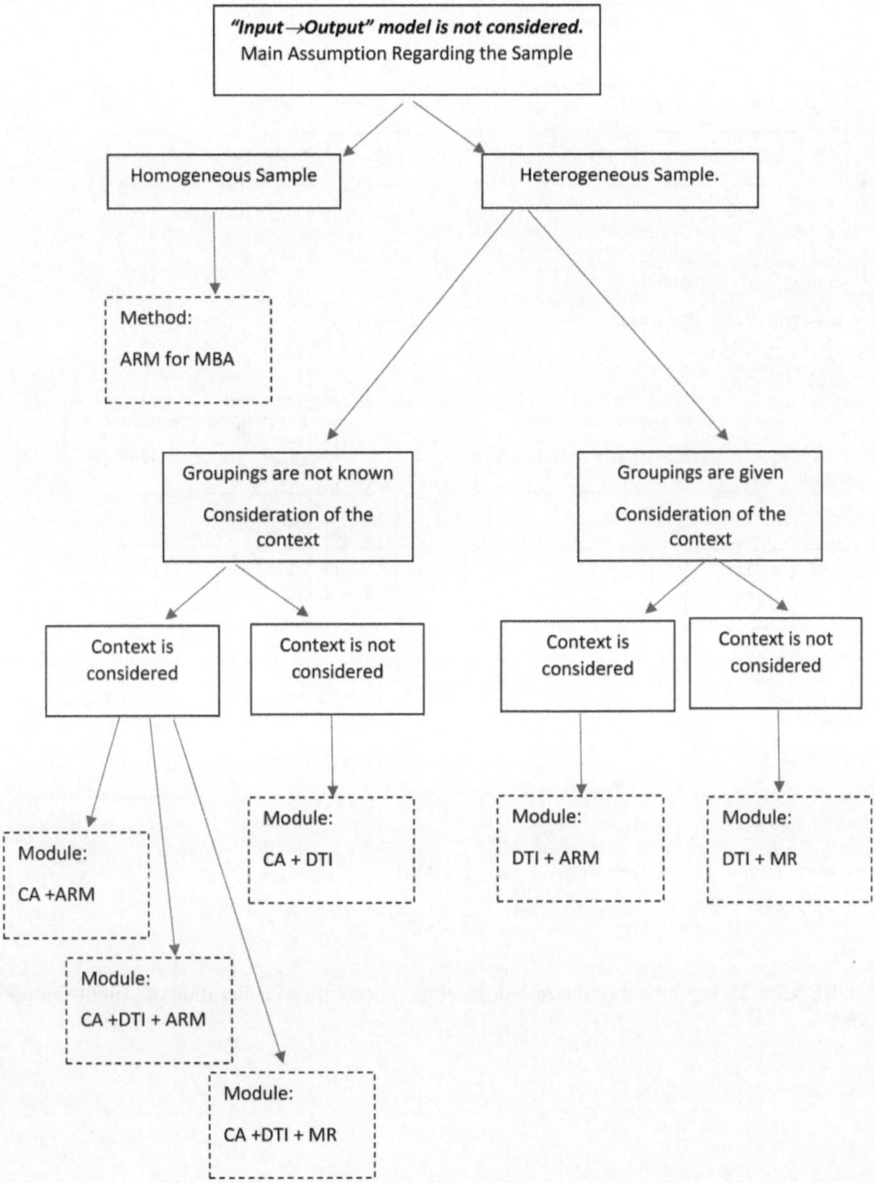


**FIGURE 3.1** Development of methodological modules without consideration of "Input-Output" model.

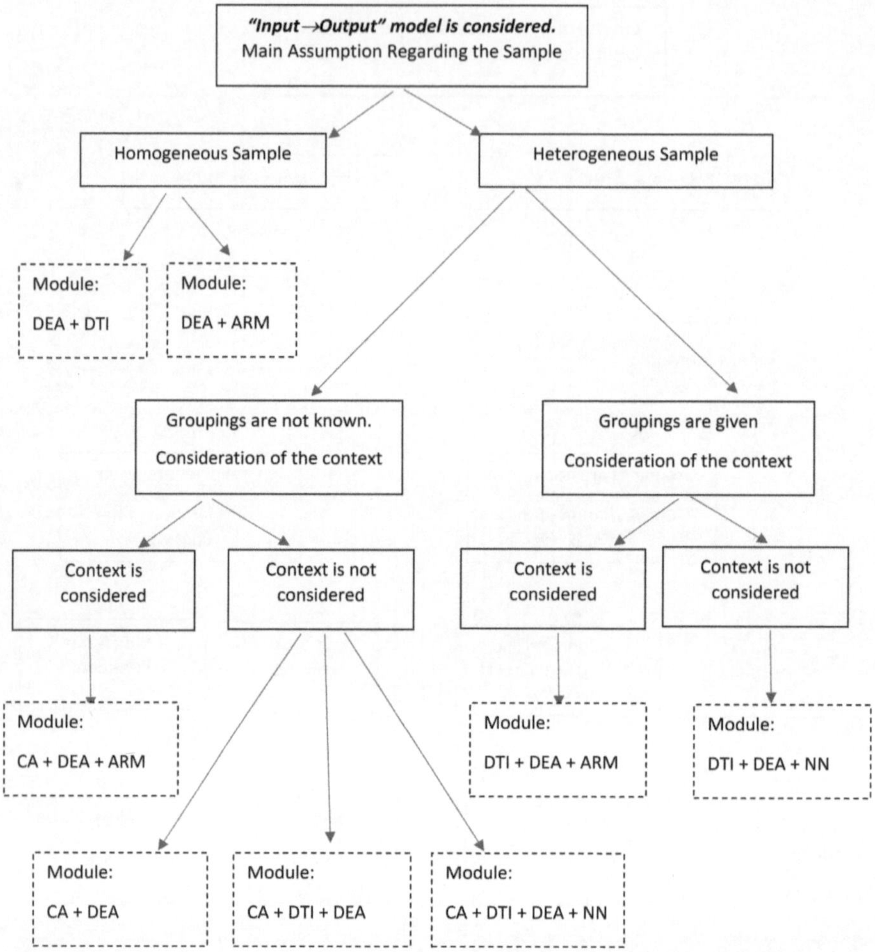


**FIGURE 3.2** Development of methodological modules with consideration of "Input-Output" model.

# *Section II*

## *Description of the Methodological Modules*

# 4 A1: Homogeneous Sample – DEA and DTI

In the case of a homogeneous sample and the presence of an "Input →Output" model, there are two basic questions that the researcher may ask:

- *What are the relatively efficient DMUs?*

and

- *What are some of the factors that differentiate relatively efficient units from relatively inefficient ones?*

This leads to a fairly simple two-phase methodology – using DEA to calculate the scores of the relative efficiency of the DMUs in the sample, and then to use DTI to identify the top-split variables.

## PHASE 1: DEA

Given the sample, the investigator must decide on the variables comprising DEA model. One of the options is to consider all the available variables to be a part of the model, while the second option is to construct the model out of a sub-set of the available variables.

Once DEA models have been specified, the investigator needs to select the orientation of the model (e.g., input, output, base) and to make a decision regarding the return-to-scale (e.g., constant, variable, non-increasing). After running the analysis, the scores of the relative efficiency will become available and the relatively efficient units will be identified.

Another decision to make is about the threshold separating relatively efficient and relatively inefficient DMUs. Strictly speaking, only DMUs with the score of 1 will be considered efficient. Using this criterion is not problematic if the sample is large and the assumption regarding the return to scale "encourages" a multi-DMU frontier (e.g., variable or non-increasing).

However, following this strict rule may result in the case where there is only one relatively efficient DMU (e.g., in the case of constant return to scale), or in the case where there is only few efficient DMUs relative to the size of the sample. In this case, a consideration could be given to selecting the efficiency threshold being lower than 1, where, for example, a cut-off score of 0.9, or 0.95 and so on could be considered. This will address the issue of having a small number of relatively efficient DMUs per sample, but, of course, would require the investigator to make a substantiating and convincing argument that defends such decision and the chosen threshold.

## PHASE 2: DTI

As a pre-requisite to running DTI, the investigators must create a new target variable that will allow for encoding relatively efficient and relatively inefficient DMUs. For all intents and purpose, a simple binary encoding, such as via assigning the value of 1 to all relatively efficient DMS and the value of 2 to the relatively inefficient DMIs, will do the job.

Another step prior to conducting the analysis is to consider the variables included in the data set. There are two options: first, to run DTI with the variables comprising the DEA model, plus the newly created target variable. This may result in the top-level splits identifying the most important differentiating variables from the DEA model.

The second option applies to the situation when the DEA model was comprised of a sub-set of all the available to the investigator variables. In such case, the researcher may consider running DTI utilizing all the available variables in the data set. This may not result in top splits based on the input or output variables of the DEA model, but it may point to the important contextual variables being responsible for the difference between efficient and inefficient DMUs.

## EXAMPLES OF APPLICATION OF DEA AND DTI

**Chapters in this book:** 18, 22, 25, and 26.

**Some published papers:**

Samoilenko, S. and Osei-Bryson, K.M. (2008). Increasing the Discriminatory Power of DEA in the Presence of the Sample Heterogeneity with Cluster Analysis and Decision Trees, *Expert Systems with Applications*, *34*(2), 1568–1581.

Samoilenko, S. (2008). Contributing Factors to Information Technology Investment Utilization in Transition Economies: An Empirical Investigation, *Information Technology for Development*, *14*(1), 52–75.

Samoilenko, S. and Osei-Bryson, K.M. (2019). A Data Analytic Benchmarking Methodology for Discovering Common Causal Structures that Describe Context-Diverse Heterogeneous Groups, *Expert Systems with Applications*, *117*, 330–344.

Samoilenko, S. and Osei-Bryson, K.M. (2013). Using Data Envelopment Analysis (DEA) for Monitoring Efficiency-Based Performance of Productivity-Driven Organizations: Design and Implementation of a Decision Support System, *OMEGA*, *41*(1), 131–142.

# 5 A2: Homogeneous Sample – DEA and ARM

In order to successfully apply this module, the investigator must have an access to a data set that contains a larger number of variables than those comprising the DEA model – we'll refer to them as *contextual* variables. For all intents and purposes, the research problem that can benefit from applying this module could be conceptualized as a "system in its environment", where the investigator is interested in the presence of the impact of the environment on the performance of the system.

In the case of a homogeneous sample there are two basic questions that the researcher may ask:

- *What are the relatively efficient DMUs?*

and

- *What are some of the contextual differences between relatively efficient from relatively inefficient units?*

This leads to another simple two-phase methodology – using DEA to calculate the scores of the relative efficiency of the DMUs in the sample, and to use ARM to generate potentially interesting itemsets.

## PHASE 1: DEA

Similar to the previous module, prior to conducting the analysis the investigator must make her decision regarding the orientation of the model and select the assumption regarding return-to-scale. Given the goal of obtaining insights into the factors impacting the level of relative efficiency of the DMUs, the decision must be made regarding the inclusion of relevant contextual variables into the DEA model.

One of the options is to consider only essential to the input–output process variables, while the second option is to also include some of the possibly important contextual variables. For example, if analyzing the efficiency of the retail stores or restaurants, the investigator may select the model that does not consider the context, or, conversely, include such contextual variables as "population density" or "household income level" in the model. Regardless of the choice of variable included in the model, it is a responsibility of the investigator to justify it.

## PHASE 2: ARM

In order to apply ARM to a numeric data set the investigator will have to perform some data conversion. In most of the cases the conversion will be based on "bucketizing" the values of each variable in the data set. For example, given a variable with the range of values from 0 to 100, we can create such obvious "buckets" as:

- "high" and "low", where "high" is assigned to the values of 50 and above, and "low" to the rest
- "high", "midhigh", "midlow", and "low", where the bucketizing is based on four quartiles
- "1ten" to "10ten", were the buckets are based on 10% increments, and so on.

Fundamentally, the only guideline for such conversion is that it has to make sense to the investigator and her audience, and to be useful for the purposes of the inquiry.

The next decision is regarding the handling of efficient vs. inefficient DMUs. One option is to separate the data set into two sub-sets and run a separate analysis for each subset. This is a viable option only in the case if the both sub-sets are sufficiently large to be analyzed. The second option is to use a complete data set to run ARM. We suggest the second option because it alleviates the demands of ARM regarding the minimum number of transactions. However, this will require creating a dummy variable that will allow for encoding the relative efficiency of each DMU. For example, a dummy variable "Efficiency", with the values "efficient" and "inefficient", will do the job of creating a logical partitioning of the data set just fine.

Because the purpose of ARM is to generate "if→then" itemsets, we can consider applying the method for the following purposes:

1. To identify itemsets comprised of only contextual variables, plus the newly created dummy variable. This, for all intents and purposes, would allow for discovering commonalities and differences in the contexts of efficient and inefficient DMUs, as well as the impact of the contextual variables on the level of relative efficiency, in the form of "If→Efficient" and "If→Inefficient" itemsets.
2. To generate itemsets comprised of the contextual variables, dummy variable, and the inputs of the DEA model. This will allow, possibly, for obtaining some insights regarding the impact of contextual variables on the inputs of the DEA model.
3. To generate itemsets comprised of the contextual variables, dummy variable, and the outputs of the DEA model. Similarly, this would allow for getting the rules showing the associations between the contextual variables and the output of the DEA model.

It is possible, to a certain extent, to manipulate the composition of the itemsets by simply excluding some of the variables from ARM analysis.

## EXAMPLES OF APPLICATION OF DEA AND ARM

**Chapters in this book:** 23 and 24.

**Some published papers:**

Samoilenko, S. and Osei-Bryson, K.M. (2019). Representation Matters: An Exploration of the Socio-Economic Impacts of ICT-Enabled Public Value in the Context of Sub-Saharan Economies, *International Journal of Information Management*, *49*, 69–85.

Samoilenko, S. and Osei-Bryson, K.M. (2020). Start a Business, Get a Credit, Make an Impact: Do ICTs Help? Impact of ICT on Legitimization of SMEs, *International Journal of Information Communication Technologies and Human Development*, *12*(2), 29–47.

# 6 B1: Heterogeneous Sample (Groupings Are Given) – DTI and ARM

This methodological module is well-suited to the situation where the investigator is interested in identifying the sources of differences between the given sub-sets of the sample, as well as in getting insights regarding the associations that describe each sub-set. The typical research questions that could be answered using this module are:

1. What are the sources of heterogeneity responsible for differentiating the sub-sets of the sample?
2. Are there associations of variables, in the form of "if→then" rules, that describe each sub-set?

## PHASE 1: DTI

The successful result of the analysis during this phase is dependent on whether the data set actually contains the variables that, indeed, differentiate the given groupings. Let us consider a scenario where the data set is comprised of the data points about HR Departments of middle-sized firms, and where the assigned groupings are based on the regions of the US. It is probable that DTI would not yield a meaningful solution, for it is not likely that the variables describing the HR Departments would be influenced by the geography. Conversely, if the researcher has an access to a data set on the economic profiles of the counties of the world, and the given groupings are based on the four categories of income level, then the results of DTI would most definitely yield a meaningful solution because the income level of a country is associated with its economic profile.

Prior to DTI, the data set must be augmented by target variable – "Group" seems to be a descriptive and simple label for such variable. The easiest way to encode the sub-sets is by assigning a simple numeric value to every sub-set, where for, let us say, four sub-sets the target variable will have domain of four values: "1", "2", "3", and "4".

Once the analysis is completed and the classification tree generated, an important decision must be made regarding the composition of the resultant nodes. Ideally, we would want to have top-level splits resulting in "pure" nodes containing 100% (or so) of a given sub-set of the sample. However, in most of the cases, the complete "purity" of a node is not obtainable. Instead, the investigator must decide on the value of what could be called a "dominance threshold" – the percentage of the node's membership for a given group that would allow to claim the dominance of the node by that group. This decision regarding the value of the dominance threshold resides under the purview of the investigator; thus, the researcher should be prepared to justify the chosen number. While 100% node's membership of a group is desirable, but likely not attainable, 60% is easier to get, but not very convincing; based on our experience, we suggest the value of at least 80%.

The number of the sub-sets in the sample plays a role in the depth of the differentiating splits. If a sample is comprised of two sub-sets, then a two-level structure of the split may produce a good result. But if the sample consists of four sub-sets, then, based on our experience, a four-level split structure is only to be expected.

One of the ways allowing to guide a search for differentiating criteria is associated with a step-by-step elimination of the top-split variables. This approach is based on running multiple DTI, starting with the complete data set, and then excluding, after each iteration, the resultant top-split variable from the data set. While this approach may not be helpful all the time, it sometimes produces valuable insights into the nature of the differences between the groups in the sample.

## PHASE 2: ARM

Prior to conducting ARM, the researcher already knows the number of the sub-sets in the sample (via received knowledge), as well as the variables playing role in differentiating the sub-sets. Consequently, there are three basic scenarios/questions that the investigator may consider pursuing in Phase 2, as follows:

1. *What are the general itemsets that characterize each sub-set of the sample?*

   In the first scenario, the investigator is interested in constructing the itemsets, for each sub-set, that are based on *all* available data. The purpose is to compare and contrast the generated rules across all the groups in the sample.
2. *What are the itemsets that characterize the sources of heterogeneity (e.g., differentiating the sub-sets variables)?*

   According to the second scenario, the investigator may want to inquire into the similarities and differences in itemsets that are associated with the top-split variables generated by DTI. For all intents and purposes, the goal is to inquire into the scenario of the type "if a variable A causes X and Y to be different (e.g., differentiates A from B), then what associations characterize A"?
3. *What are specific itemsets, comprised of differentiating the sub-sets variables, which characterize each sub-set of the sample?*

The third scenario is utilized when the investigator wants to confirm whether the differentiating variables identified by the top-split variables of DTI are associated (e.g., are a part of *If→Then* itemset) together. While the results of DTI produce the top-split variables, the nature of the decision tree induction does not allow for insights into the possible associations between the split variables. However, if associations between the differentiating variables are found, then it may lead to further inquiries regarding possible synergies and/or complementarities between the top-split variables.

In order to answer the stated above questions, the investigator must be able to identify each transaction (e.g., record) as belonging to one of the sub-sets of the sample. In the simplest case, the data set is large enough to be partitioned into the sub-sets, where each sub-set could be subjected to ARM. In such situation, generated itemsets could be compared and contrasted across the sub-sets of the sample.

Most of the time, however, the opportunity of running a separate ARM for each sub-set is not there, simply because of the size of the data set. There are two options that could be considered. First, in the case where each sub-set is characterized by the different level(s) of the variables of interest, for example, when dealing with the partitioning of the sample based on the criteria of income level of economy, the sub-sets could be "low income", "middle income", and "high income" economies. And in the case of, let us suppose, economic data, each economy will be represented by such variables as "GDP" and "GDP per capita". In such situation, clearly, "low income" country would have values "low GDP" and "low GDP per capita", "middle income" would have "middle GDP" and "middle GDP per capita" and so on. In this scenario, basically, each sub-set could be identified by the level of the variables describing the sub-set.

The second option is to be considered where each sub-set has overlapping with other sub-sets range of variables describing it. In such case, the investigator would have a "mixed bag" of values for each variable describing each sub-set, where it would not be possible to identify each sub-set by the values of the variables that represent it. In this situation, the investigator may consider creating a dummy variable, let us say, "Group", with the values identifying each sub-set, such as "Group1" and "Group2".

When the investigator is ready to conduct ARM, we suggest she proceeds in a hierarchical top-to-bottom fashion, starting from the complete set of the available variables, and gradually excluding the variables from the analysis. This should allow for pursuing the three scenarios we outlined above.

## EXAMPLES OF APPLICATION OF DTI AND ARM

**Chapters in this book:** 23 and 24.

**Some published papers:**

Samoilenko, S. and Osei-Bryson, K.M. (2019). Representation Matters: An Exploration of the Socio-Economic Impacts of ICT-Enabled Public Value in the Context of Sub-Saharan Economies, *International Journal of Information Management*, *49*, 69–85.

Samoilenko, S. and Osei-Bryson, K.M. (2020). Start a Business, Get a Credit, Make an Impact: Do ICTs Help? Impact of ICT on Legitimization of SMEs, *International Journal of Information Communication Technologies and Human Development*, *12* (2).

# 7 B2: Heterogeneous Sample (Groupings Are Given) – DTI and MR

This methodological module is well-suited to the situation where the investigator is interested not only in identifying the sources of heterogeneity of the sub-sets of the sample but also in knowing whether the received categorization is associated with differentiating factors. The typical research questions that could be answered using this module are:

1. What are the factors differentiating the pre-defined sub-groups in the sample?
2. What is the impact of differentiating variables on a common causal model of the pre-defined sub-groups?

Let us consider a simple illustrative example for such methodological module. Let us say, we have a data set representing three groups of economies: low income, middle income, and high income. The classification based on the income level represents a received knowledge provided by the taxonomy of the International Monetary Fund. The set of available variables reflects various aspects of an economy, such as infrastructure, labor market, variety of economic indicators, and so on. Furthermore, we have a causal model (e.g., the framework of Neoclassical Growth Accounting) to work with, and according to the model the sources of economic growth (dependent variable) are capital (independent variable), labor force (independent variable), and changes in productivity (error term).

Given such scenario, the questions that could be of interest answering are:

1. What are the factors associated with the received heterogeneity of the sample? Meaning, other than the label assigned to each of the groups, what are the factors that actually differentiate the groups?
2. Given the common causal model, what is the impact of the differentiating variables on the causal model? Also, do differentiating variables associate with independent, or dependent variables of the common causal model?

Such questions are important to answer, for what if, given the common causal model of "Economic growth = (Capital, Labor, TFP)", Capital and Labor would end up being statistically significant for high income economies, only Capital for middle income economies, and neither Capital, nor Labor would be statistically significant for low income economies? The follow-up question would be: "what are the reasons for such discrepancy"?

## PHASE 1: DTI

Performing DTI in this module is not very different from the previous examples. The important consideration is regarding the received knowledge about the causal model that is used in Phase 2. Let us consider the most obvious scenarios of this phase- the questions that the investigator may pose are:

1. Is there an association between the received categorization and the causal model?
2. Is there an association between the received categorization and the contextual variables?
3. Is there an association, based on the received categorization, between the contextual variables and the causal model?

### Option 1: DTI Using the Data Set Comprised of a Causal Model Only

In the case of a large data set and a known causal model, the researcher has an option of performing DTI using the data set comprised only of the variables of the causal model, plus, the target variable differentiating the sub-groups. This may result in top-level splits being based on the variables of the causal model-independent variables or dependent variable. After running DTI, the investigator will have enough information to determine if the categories are associated with one or more variables of the causal model.

### Option 2: DTI Using the Data Set without Causal Model

In this option, DTI is performed on the data set without including the variables that are a part of the casual model. This will result in possible identification of the contextual variables that play role in differentiating the sub-groups and which are, possibly, related to the received categorization. And such contextual variables may, indirectly, be associated with the variables of the causal model.

### Option 3: DTI Using the Complete Data Set

The last option of DTI uses a complete data set comprising a causal model and contextual variables. While similar in nature to two previous options, the benefit of running this model is that it may allow for identifying a set of variables that play role in differentiating the data set based on the received categorization. By eliminating, in a step-by-step fashion, the top-split variables from the data set and re-running the analysis, the investigator has an opportunity of identifying a hierarchy of variables differentiating the categories.

## PHASE 2: MR

The purpose of the second phase of this module is to test the common causal model across the sub-sets of the sample. We assume that the sub-groups are large enough and we have a sufficient number of degrees of freedom to run MR for each sub-group.

## Option 1: MR Using the Causal Model Only

The usual approach is, of course, to test MR model including the interaction terms, so, we will not spend any time covering this point. So, the simplest way of proceeding with Phase 2 is to run MR, based on the common model, for each of the subgroups of the sample. This option is a preferred one if DTI of Phase 1 did not yield any interesting top-level splits that may impact the common causal model.

## Option 2: MR Using the Adapted Causal Model – Contextual Independent Variable

Let us assume that Option 2 of DTI resulted in top-level splits that are not based on the variables of the causal model. This gives us an insight that the identified variables could play a role in the differences (if there is a difference to be discovered) of the results of testing the causal model in the context of the sub-samples of the data set. So, the investigator could select augmenting the common causal model by including one or more independent variables discovered by the results of DTI. The caveat is, of course, that such augmentation of the model has to make sense to the investigator and has to be defensible when presented to the audience.

## Option 3: Creating a New MR Using Contextual Independent Variables

Sometimes, the results of DTI produce the top-level splits that are based on the variables that appear to be predictors (or, could be associated with) of the independent variable(s) of the common causal model. For example, let us suppose that while using our example of the common causal model of "Economic growth = (Capital, Labor, TFP)", the results of DTI identified the "healthcare expenditures" and "education expenditures" being the top-split variables differentiating the sub-samples. If the investigator can make a solid defensible argument for a presence of a new "independent variables → dependent variable" model, then the obtained insights may result in creating another causal model, to augment the existing one.

As a result, it is not inconceivable to consider a model of the type "Labor = (*Healthcare Expenditures, Education Expenditures*)" to be useful, for a solid argument could be constructed that such quality of life expenditures as healthcare and education should have an impact on the labor force of an economy. While, undeniably, an addition of an extra MR model requires an extra effort on the part of the investigator, the benefits could be worthy, for it may significantly expand a chain of links comprising the original research model.

# EXAMPLE OF APPLICATION OF DTI AND MR

**Chapter in this book:** 18.

**Published paper:**

Samoilenko, S. and Osei-Bryson, K.M. (2013). Using Data Envelopment Analysis (DEA) for Monitoring Efficiency-Based Performance of Productivity-Driven Organizations: Design and Implementation of a Decision Support System, *OMEGA*, *41*(1), 131–142.

# 8 B3: Heterogeneous Sample (Groupings Are Given) – DTI, DEA, and ARM

This is a more complex module, and, unsurprisingly, it allows for answering a set of more interesting questions. The basic setting for using the module is a data set comprising multiple sub-sets that are described by a common DEA model. This module allows the investigator to answer the following questions:

1. What are the differentiating variables that are associated with the received categorization? Answering this question allows for identifying the dimensions that separate the groups.
2. Do differentiating variables comprise a part of the DEA model? Answering this question allows for discovering whether the differences between the sub-groups are due to the differences in inputs or/and outputs that will be used in calculating the scores of relative efficiency.
3. Is there an association between the levels of inputs and outputs of the DEA model and the sub-groups of the sample? While previous question helps finding out whether input/output variables of the DEA model are associated with the received categorization, this question deals with the *levels* of differentiating variables. For example, given a categorization of high, middle, and low income economies, we could identify "capital" as being a variable of the top split. However, it is not clear whether the received categorization maps on the levels of "capital" (e.g., high income economies have "high capital" and middle income economies have "mid capital").
4. What is the level of relative efficiency of each sub-set in the group? While we may expect that all the sub-groups contain the relatively efficient and relatively inefficient DMUs, we would not expect that, on average, the sub-groups would fare equally in terms of the scores *per sub-group*.
5. Is there an association between the differentiating variables and the averaged scores of the relative efficiency of each sub-group? Fundamentally, we would like to identify the sources of the disparity of the averaged scores of the relative efficiency between the sub-groups. While we know, at this point, that there is a difference between the sub-groups in terms of the top-split variable(s) identified by DTI, we don't know whether the two "differences" are related.

6. Is there an association between the DEA model and the itemsets generated by ARM? In order to inquire into the nature of the differences between the sub-groups in terms of the scores of the relative efficiency, it could be useful to identify interesting associations including (1) inputs of DEA model, (2) outputs of DEA model, and (3) scores of the relative efficiency.

For the illustrative purposes, let us consider an example of a data set representing three groups of economies: low income, middle income, and high income. Let us also consider a common DEA model comprising the two input variables, namely, "Investments in ICT" and "ICT Labor Force" and a single output variable "GDP". We would like to find answers to two big questions, namely:

1. Is there a difference between the groups of economies in terms of the averaged scores of the relative efficiency of conversion of "Investments in ICT" and "ICT Labor Force" into "GDP?"
2. What are some of the factors that could be responsible for the differences?

## PHASE 1: DTI

Phase 1 of this module allows for answering questions 1, 2, and 3 from the list above. Prior to performing DTI, we need to consider whether the data set is comprised only of the variables constituting the DEA model, or, if the data set also contains contextual variables.

### Option 1: The Data Set Is Comprised of the Variables of the DEA Model

In this case, the results of DTI produce top level splits indicating the difference in terms of the inputs and/or outputs of the DEA model. In the simplest case, the received categorization will be supported by the corresponding splits, so we'll get the results of the type where high income economies would have high levels of *ICT Labor Force* and *Investments in ICT* and/or *GDP*, and so on. However, in some cases the received categorization will not be supported by a clear-cut delineation of the sample based on the top-level splits. In any case, the results of DTI in the case of this option are not to be expected to be very telling – at the best we could discover that the sub-groups of the sample differ in terms of the levels of the inputs and outputs of the DEA model.

### Option 2: The Data Set Contains Contextual Variables

In the case of the second option, the results could vary greatly and will be impacted by the number of the available variables. Ideally, by the process of continuous elimination of the top-split variables from the analysis, the investigator will be able to identify a set of the variables that differentiate, clearly, the sub-groups of the sample.

Overall, the first phase of the analysis is the crucial one to the goal of the inquiry, for if we cannot identify a set of variables that clearly (e.g., based on the chosen by the investigator' threshold for the purity of the node) differentiate the sub-groups, then the inquiry will be reduced to a simple calculation of the averaged DEA scores for each group.

## PHASE 2: DEA

Phase 2 of this module allows for answering the question 4 from the list. The analysis of this phase is a straight-forward affair, with one caveat. Once the investigator settled on the choice of the model and the assumption regarding return-to-scale, the scores of the relative efficiency could be calculated. Commonly, the analysis will be performed on the complete data set relying on logical partitioning of the sample. In such case, every DMU will be associated with its category within the sample, and once the scores have been generated, the results will be grouped according to the categories. Averaging of the scores for each sub-group will offer an opportunity to compare the groups in terms of their averaged relative efficiency of the conversion of inputs into outputs.

In some cases, it is useful to physically partition the sample into the sub-groups and run the second DEA for each of the groups separately. This will allow the investigator to identify the best performers in each sub-group. The benefit of the second analysis is clear in the case where benchmarking of the best performers is considered. Let us suppose that we identified a poor-performing low income economy *A*, and we also identified a high-performing middle income economy *B*. It could be a tall order to attempt to benchmark the performance of *B* by *A*. Instead, it could be a better strategy to identify, first, a set of high-performing low income economies, so one of them could be benchmarked by *A*.

This opens an opportunity for a hierarchical benchmarking, where a poor performer emulates the better performing peer within its group, and after the performance has been improved, attempts emulating a better performer from a different group.

If the investigator has an access to a time-series data, then a possibility of conducting MI should also be considered. This would allow for identifying the changes in relative efficiency over time, as well as inquiring into the sources of changes.

## PHASE 3: ARM

The last phase of this module allows for answering questions 5 and 6 of the list we provided in the beginning. Realistically, this phase makes sense to conduct only if the investigator has a data set with contextual variables in addition to the DEA model' data. Thus, we consider the availability of such "expanded" data set to the investigator a given.

Let us recall that we arrived to Phase 3 with the knowledge of the differentiating the sub-groups in the sample' factors and with the calculated averaged scores of the relative efficiency for each of the sub-group. In general, the results of the first two phases are insufficient to produce some recommendations regarding how to improve the scores of the relative efficiency of the lower performing DMUs in the sample. This is because, based on the DEA model, all recommendations will be based on manipulating the values of inputs and outputs. If we consider a "(Capital, Labor) →GDP" DEA model, then in order to improve the scores of the relative efficiency based on the input orientation, we would have to suggest lowering the levels of the inputs: capital and/or labor. Intuitively, lowering the quantity of labor force via increasing its quality (e.g., education and training) is an attractive proposition, but the question is whether it could be done. Simply put, it is important to know why the quantity of labor for a given economy is what it is. Similarly, another option would be to decrease the amount of allocated capital (e.g., investments) in order to do "more with less". However, such recommendation is simplistic, for, again, we don't know the underlying associations that could be important to the amount of allocated capital. This becomes even more important when a decision maker cannot impact the variables of the DEA model directly, but, instead, may only have an indirect influence on them. Again, the quantity of the labor force is a good example, for a correct way of reducing the number of workers is not simply by attrition, but by making the workers more efficient and effective, which requires implementation of a variety of workforce development programs. Simply put, we would like to get some insights into the context within which each sub-group exists.

In this situation, undoubtedly, we would like obtain a set of additional insights into the nature of the contextual differences between the groups. Specifically, we may aim to discover some interesting itemsets reflecting:

1. The differences between the sub-groups in the sample. For example, it could be useful to discover the association rules of the type "If (var1, var2, var3, etc.) → then (top-level split' differentiating variable)". If there is a factor, or a set of factors, that differentiate the sub-groups, then what are the associations that play role in a sub-group having a particular level of a differentiating variable?
2. The levels of the input variable of the DEA model. If, for example, low income economies have a low level of labor, then what are the associated factors?
3. The levels of the output variables of the DEA model. If high income economies have a high level of GDP, then what are the associations of the type "If (var1, var2, var3…) → Then (GDP)" that may exist?
4. Additionally, we may be interested in identifying the itemsets that characterize the levels of averaged relative efficiency. For example, if low income economies ended up having "low" averaged scores of the relative efficiency, then what are the factors that are associated with it?
5. Finally, we may want to discover associations that describe the groups in terms of the received categorization. For example, if we have low income, middle income, and high income groups of economies, then this option offers us an opportunity to describe each category in terms of the itemsets based on the available variables.

This leads us to pursuing five ARM scenarios, which we describe below. Despite presenting the scenarios in term of options, we suggest to our reader using them together, in a step-by-step fashion.

### Option 1: ARM to Generate "If→ (Level of the Top-Split Variable(s))"

For the purposes of this option, it is useful to remove from the data set all the variables that comprise the DEA model – this will prevent from generating self-referential association rules. The results would yield the itemsets associated with the specific levels of the top-split variables and would allow for answering the question "what associations of the factors describe the sub-groups in the sample?"

### Option 2: ARM to Generate "If→ (DEA Model's Inputs)"

For the purposes of this option, it is beneficial to remove from the data set all the variables that comprise the outputs of the DEA model. Using this option, we would like to find out what itemsets characterize the levels of the inputs of the DEA model. This is important, because the levels of the inputs may not be similar to the levels of the top-split variables, or to the received categorization (e.g., low income economy may have high level of labor as one of the DEA inputs). By employing this option, the investigator should be able to answer the question "what characterizes the certain level of inputs of the DEA model"?

### Option 3: ARM to Generate "If→ (DEA Model' s Outputs)"

This option gives the investigator an opportunity to discover whether the levels of the output variables(s) of the DEA model are associated with the levels of the inputs. We would expect the presence of the parity of levels, where, for example, DMUs with "low" levels of inputs would also have "low" levels of outputs. This option allows us for discovering interesting deviations from what is expected.

### Option 4: ARM to Generate "If→ (Level of Averaged Relative Efficiency)"

For the purposes of this option, it will be required to create a new variable representing the level of averaged relative efficiency representing each sub-group in the sample. If, for example, the results of DEA yielded a three-level set of scores (e.g., low income economies got "high" average and high income economies got "low" average), then a variable "Efficiency", with the domain of values "low", "medium", and "high" would do the job.

### Option 5: ARM to Generate "If→ (Received Categorization)"

For the purposes of this option, the investigator may consider using a complete data set. In this option, we are interested to find out the itemsets describing the sub-sets of the sample. This option is especially useful to apply in the case if a received categorization is not level-based. For example, a designation of low income, middle income, and high income economies is based, clearly, on the level of economic well-being. Thus, we would expect to have some correspondence in the levels of the variables comprising itemsets that describe each category to that category (e.g., low income economy having low level of GDP). However, if the received categorization is nominal in nature, then the level-based correspondence would not apply. Let us consider a designation based on a geographical location instead of the level of economic strength, such as Sub-Saharan, Middle Eastern, Eastern European, and Latin American economies. By applying ARM, we can, possibly, generate interesting associations characterizing each group in the sample.

## EXAMPLES OF APPLICATION OF DTI, DEA, AND ARM

**Chapters in this book:** 23 and 24.

**Some published papers:**

Samoilenko, S. and Osei-Bryson, K.M. (2019). Representation Matters: An Exploration of the Socio-Economic Impacts of ICT-Enabled Public Value in the Context of Sub-Saharan Economies, *International Journal of Information Management*, *49*, 69–85.

Samoilenko, S. and Osei-Bryson, K.M. (2020). Start a Business, Get a Credit, Make an Impact: Do ICTs Help? Impact of ICT on Legitimization of SMEs, *International Journal of Information Communication Technologies and Human Development*, *12*(2), 29–47.

# 9 B4: Heterogeneous Sample (Groupings Are Given) – DTI, DEA, and NN

This module allows the researcher to answer questions regarding the underlying causes of the differences in the scores of the relative efficiency of the sub-groups in the sample. Specifically, the differentiation is made between the culprit being (1) the differences in the levels of inputs/outputs or (2) the transformative capacity – capability of DMUs to convert inputs into outputs. An important requirement for this module is a size of the sample available to the investigator – it has to be large enough to allow for running NN for each of the sub-groups of the sample. In terms of the composition of the data set, the sample is comprised of the variables representing inputs and outputs of the DEA model – contextual variables, for all intents and purposes, are not considered for this module.

Overall, this methodology allows for answering the following research questions:

1. What are the sources of the heterogeneity of the sub-groups of the sample? (DTI)
2. Do sources of heterogeneity represent input or output variables of the DEA model? (DTI and DEA)
3. What are the scores/levels of the relative efficiency of the DMUs in each of the sub-groups of the sample? (DEA)
4. Is there a difference between averaged scores of the relative efficiency of the sub-groups of the sample? (DEA)
5. What are the sources of the differences in averaged scores of the relative efficiency between the sub-groups of the sample? (NN and DEA)

For illustrative purposes, let us consider a simple scenario of having a historic data about two sales teams, Team A and Team B, described by the DEA model (# of members, # of locations → # orders placed). Two reasonable questions that we may have are:

1. In what way Team A and Team B are different?
2. Which team is more efficient?
3. What are the reasons for the inefficiency of a less efficient team?

## PHASE 1: DTI

Prior to conducting DTI, the investigator needs to create a target variable differentiating the sub-groups of the sample (e.g., a variable "Group" with the domain of values "A, B"). The results of the first phase provide an insight into whether the differences between the sub-groups in the sample (e.g., Team A and Team B) are associated with the inputs and/or outputs of the DEA model.

It is quite possible that the results will not produce any relatively clean-cut top-level splits, meaning that the sub-groups do not differ, significantly, in terms of the inputs and/or outputs of the DEA model. If this is the case, however, then it is not necessarily true that the sub-sets of the sample would not differ in terms of the average values of the scores of the relative efficiency.

On the other hand, it could be that the top-level splits would indicate that the sub-groups do differ in terms of the variables comprising the DEA model. Similarly, this does not indicate that the sub-sets of the sample would significantly differ in terms of the scores of the relative efficiency. Simply put, the most important insight offered by DTI here is whether the sub-groups of the sample are "truly" different in terms of the variables comprising the DEA model. This determination will become important after Phase 2 is completed.

## PHASE 2: DEA

Conducting DEA at this phase is a straightforward affair consisting of selecting the orientation of the model and selecting the assumption regarding return to scale. No physical partitioning of the sample is required, so the DEA is conducted using the complete data set. Once the scores have been obtained, DMUs are grouped according their membership in the sub-set they belong to, and the scores for each of the sub-sets are averaged. It is expected that the result of the analysis would show a difference in the averaged scores of the relative efficiency between the sub-sets (again, the threshold of the differentiation is under the purview of the investigator). However, DTI and DEA are not sufficient to answer the question regarding the nature of the differences.

Using our example of a Team A and a Team B, and assuming that Team A has a higher averaged scores (for input-oriented model) of the relative efficiency than Team B, we are not in a position of answering the question *Why Team A is relatively more efficient than Team B?* This is because the difference could be due to two factors:

1. Difference in the levels of the inputs and outputs of the DEA model (if this is supported by the results of Phase 1), and
2. Difference in the transformative capacity between the teams, where Team A is better at converting inputs into outputs (for input-oriented model) or require less inputs to produce the same level of outputs (for output-oriented model) than Team B.

While it is important question to answer from the perspective of the research, the practical significance is also very clear. Let us consider that we want to improve the level of efficiency of Team B. If we assume that the difference in the level of performance is due to the level of the inputs, then we'll add more people and more locations to Team B, hoping that this would result in the increased number of orders. However, it is possible that by doing so we would waste the resources and the performance of Team B (e.g., number of placed orders) will, instead, deteriorate. This is because it is also very possible that Team B simply is not as good as Team A – in this situation the proper avenue for improving the performance of Team B is via team development and improvement of the required skills. So, by obtaining insights into the nature of inefficiencies we could better construct the intervention program allowing for the most effective allocation of resources.

## PHASE 3: NN

The purpose of this phase is to create what we call a *model of transformative capacity* for each group in the sample. For all intents and purposes, it amounts to training NN to develop the weights of the hidden nodes for each group, in the case of our running example, Team A and Team B. This phase proceeds in a four-step fashion.

### Step 1: Generate NN Model of Transformative Capacity

First, we ran NN for Team A (we leave the details of training and testing of the NN outside the scope of our discussion here) with the purpose of saving the generated NN model. Second, we do the same for Team B. The outcome of this step is saved as *n*-models of NN, based on *n* sub-groups of the sample.

### Step 2: Generate Outputs of a Less Efficient Group Based on Transformative Capacity of a More Efficient Group

The purpose of this step is to generate the outputs of the DEA model of a less efficient group based on the saved NN model of the more efficient group. In the context of our example, this allows us to simulate the outputs of Team B, using their own inputs, but the process of transformation of Team A. Simply put, this allows for answering two questions:

1. What the level of outputs of Team B would be if they use their own inputs, but more efficient process of transformative capacity of Team A?
2. What the level of outputs of Team A would be if they use their own process of transformative capacity applied to the lower level of inputs of Team B?

### Step 3: Generate Outputs of a More Efficient Group Based on Transformative Capacity of a Less Efficient Group

The purpose of the last step is similar to the previous one – we would like to simulate the outputs of one group based on the model of transformative capacity of another group. In this step, we use the inputs of a more efficient sub-group to generate the outputs based on the model of transformative capacity of less efficient sub-group.

Again, keeping in mind our example of two teams, this step allows for answering two questions:

1. What the level of outputs of Team A would be if they use their own inputs, but a less efficient process of transformative capacity of Team B?
2. What the level of outputs of Team B would be if they use their own process of transformative capacity, but the higher level of inputs of Team A?

### Step 4: Compile the Generated Outputs in a New Data Set

The last step of the phase is to put together the results into the "original inputs-simulated output" data set. We should end up with two sub-sets:

1. "original inputs of less efficient group → simulated outputs based on more efficient group"
2. "original inputs of more efficient group → simulated outputs based on less efficient group"

Once it is done, we are ready to conduct the last phase of this module – DEA.

## PHASE 4: DEA

In the last phase, we perform DEA using the simulated data. During this phase, we run DEA twice.

First, we use the original sub-set of the data of the more efficient sub-group of the sample, plus, the simulated data set for the less efficient sub-group (e.g., "original inputs of less efficient group → simulated outputs based on more efficient group") that was based on the more efficient process of transformative capacity. Once the analysis is complete, we re-calculate the averaged scores of the relative efficiency of each sub-group.

Second, we use the original sub-set of the more efficient sub-group combined with the simulated data set for the less efficient sub-groups that was based on their own process of transformative capacity, but using the inputs of a better performing sub-group. Similar to the first analysis, we re-calculate the averaged scores of the relative efficiency for both sub-groups.

By comparing the averaged values of the relative efficiency of the less-efficient sub-group for both cases, we would be able to identify if the positive changes in the averaged scores greater in the case of improving the process of transformative capacity, or if the changes are greater in the case of the increasing the levels of the inputs.

## EXAMPLE OF APPLICATION OF DTI, DEA, AND NN

**Chapter in this book:** 18.
**Published paper:**

Samoilenko, S. and Osei-Bryson, K.M. (2013). Using Data Envelopment Analysis (DEA) for Monitoring Efficiency-Based Performance of Productivity-Driven Organizations: Design and Implementation of a Decision Support System, *OMEGA*, *41*(1), 131–142.

# 10 C1: Heterogeneous Sample (Groupings Are Not Known) – CA and DTI

The purpose of this simple module is to inquire into the possible heterogeneity of the sample and to identify some of the factors differentiating the sub-groups in the sample. The applicability of the module is wide – it could be used in pretty much any setting where the investigator suspects that the assumption of homogeneity of the sample may not hold. In terms of the number of variables comprising the data set, the investigator would benefit from having as much of the relevant attributes as possible. As an illustrative example for this module, we consider a data set representing a group of developing economies for a period of five years – let us say, 2006–2010. This time period should offer a glance at the economic situation before and after economic crisis of 2008.

This module would allow for answering the following questions:

1. Do developing economies in the sample represent a homogeneous group? By answering this question, we would find out if the sample is comprised of multiple sub-sets of economies.
2. Do years of representation of developing economies represent homogeneous periods? This question allows for identifying the presence of the differences in terms of years of the period. For example, it is possible that given the five-year period years 1, 2, and 3 are different from years 4 and 5. This would allow for identifying the presence of sub-periods within the time-series data.
3. If the economies in the sample, or years of representation of a period, are different, then what factors are responsible for the differences? This would allow for identifying the common variables shared by each of the sub-groups of the sample, the variables differentiating the sub-groups, as well as the variables differentiating sub-periods.

## PHASE 1: CA

The process of running CA is a straightforward affair, where the special attention should be paid to the analysis of the results. Most of the applications offer an option of examining the visual representation of the analysis, where the researcher could evaluate the relative placement and the location of the resultant clusters. While no specific guidelines would work for all cases, the general rule of thumb is the solution with the multiple clusters of a sufficient size each, and within a comparable distance from each other.

Considering an example of the time-series data representing the sample of developing economies, if the results yielded a three-cluster solution, then the researcher should investigate:

1. The membership of each cluster in terms of economies
2. The membership of each cluster in terms of the years of the time period
3. The membership of each cluster in terms of the years and economies.

In order to simplify the evaluation of the results of CA in the case of time-series data, the investigator may consider creating a three-variable scheme prior to running the analysis, as follows: "LabelYear", "Label", and "Year". For example, in the case of developing economies, where we would have "Mexico2006", "Mexico2007" and so on, it would be, for each year, "Mexico2006", "Mexico", "2006", "Mexico2007", "Mexico", "2007", and so on. This should allow for a much easier assessment of the membership of the clusters in terms of the relevant dimensions.

The expected result of Phase 1 is an n-cluster solution, where each cluster could be characterized in terms of the sub-set of the sample based on a label, and/or a year, and/or a label and a year. It is important to note that the result of Phase 1 might well be a *set* of n-cluster solutions, because, if using k-means clustering, the investigator could influence the results of the analysis if he/she wants to emphasize the cluster membership being based on the label (e.g., "Mexico", "Russia"), or a year (e.g., "2006", "2007"), or both (e.g., "Mexico2006", "Russia2006").

## PHASE 2: DTI

Prior to running DTI, the investigator should create a target variable(s) that identifies each cluster. It is quite possible that a single clustering solution may provide an opportunity for running multiple DTI analyses. This is because every cluster of n-cluster solution may contain a slightly different combination of the records based on "LabelYear", "Label", and "Year".

In the case when the investigator is in possession of the large data set comprising many possibly relevant variables/indicators, it could be beneficial to run multiple analyses. This is done in a step-by-step fashion, through the process of continuous elimination of the top-split variable from the analysis. Such approach may allow for discovering interesting differentiating factors which, otherwise, may remain unidentified in the case of a single analysis.

Finally, in the case of a multi-dimensional analysis (e.g., "LabelYear", "Label", and "Year"), it is useful to compare and contrast generated English rules for each dimension – this option offers a convenient opportunity to assess similarities and differences of the dimensions.

## EXAMPLES OF APPLICATION OF CA AND DTI

**Chapters in this book:** 18, 25, and 26.

**Some published papers:**

Samoilenko, S. and Osei-Bryson, K.M. (2008). Increasing the Discriminatory Power of DEA in the Presence of the Sample Heterogeneity with Cluster Analysis and Decision Trees, *Expert Systems with Applications*, *34*(2), 1568–1581.

Samoilenko, S. (2008). Contributing Factors to Information Technology Investment Utilization in Transition Economies: An Empirical Investigation, *Information Technology for Development*, *14*(1), 52–75.

Samoilenko, S. and Osei-Bryson, K.M. (2013). Using Data Envelopment Analysis (DEA) for Monitoring Efficiency-Based Performance of Productivity-Driven Organizations: Design and Implementation of a Decision Support System, *OMEGA*, *41*(1), 131–142.

# 11 C2: Heterogeneous Sample (Groupings Are Not Known) – CA and ARM

One of the shortcomings of the previous module is its intent to describe the heterogeneous sub-groups of the sample only in terms of the differences vis-à-vis each other. However, in many cases, it is useful to describe the sub-groups in broader terms – using differentiating as well as common dimensions. Let us consider an example of fast-food restaurants of the same type (e.g., McDonald's and Burger King) across the continental USA. The previous module would allow us to identify the presence of the sub-sets of the restaurants, and to identify the differentiating characteristics, but it would not allow to describe the sub-sets via multiple dimensions, some of which are common, and some of which are different. Using CA in conjunction with ARM allows doing so.

This module aims for answering the following questions:

1. Do entities in the sample represent a homogeneous set?
2. In the case of the heterogeneous sample, what are the descriptive characteristics of each sub-set?

## PHASE 1: CA

Let us consider two scenarios regarding the composition of the data set. In the simplest, and the most, limiting case, the data set comprises what could be called "intrinsic" variables depicting a restaurant, such as number of staff, number of dine-in customers, number of drive-through customers, and daily sales. Such a data set consists of "actionable" variables – the dimensions that each restaurant should have an ability to influence one way or another.

In the second case, the data set comprises intrinsic and environmental variables – in addition to the variables that describe each restaurant, the investigator is in possession of the variables that depict the context within which each restaurant operates. Unlike actionable intrinsic variables, contextual variables are not under control of a restaurant – given a location of a fast-food restaurant, the management cannot do anything about socio-economic status of the area or its demographic status. Given the second case, the researcher must consider which of the following three options to pursue:

1. Run CA based on only actionable intrinsic variables
2. Run CA based on only contextual/environmental variables
3. Run CA based on the complete data set – using intrinsic and contextual variables.

The three options above are not mutually exclusive – they do not preclude the investigator from running multiple analyses – one, for example, using option one, and the second using option two. The choice is, of course, dependent on the purpose of the inquiry – if the goal is the improvement in efficiency and effectiveness of the performance of lowest-tier restaurants, then running CA based on a set comprising intrinsic variables is preferable. However, if the purpose is to identify new locations for the restaurants, or to customize the menu, then the option based on the contextual variables could be a better choice.

Regardless of the selection of the variables to be used in the analysis, the expected result is an *n*-cluster solution.

## PHASE 2: ARM

The purpose of the second phase of this module is to generate a variety of itemsets, where each itemset serves its own purpose, yet is complementary to other itemsets. Let us consider some of the options using the example of a set of fast-food restaurants.

### OPTION 1: ARM USING ONLY INTRINSIC VARIABLES

This option offers an opportunity to generate rules that could be useful for the purposes of gaining insights into the internal operations of the analyzed entities. First, the investigator may decide to run the analysis using a complete data set – this may allow for confirming the results of cluster analysis via identifying level-specific rules in the data set. Let us suppose that the results of CA yielded a three-cluster solution partitioning the sample into "high-", "average-", and "low-"performing restaurants. If we are in a possession of the data regarding the average level of experience of employees and the average level of revenue per employee, then it would not be surprising if the results of ARM will generate itemsets of the type "HighExperience→HighRevenue", "MidExperience→MidRevenue", and "LowExperience→LowRevenue" for, correspondingly, "high", "average", and "low" tiers of restaurants.

After applying ARM to generate itemsets based on the complete sample, the investigator may consider running the analysis based on the sub-sets of the sample – using the clusters identified in Phase 1. This approach allows for generating itemsets describing each of the sub-sets much more specifically. Once the itemsets were generated for each cluster, the resultant "If→Then" rules could be easily compared and contrasted vis-à-vis the other sub-sets of the sample.

Overall, the benefit of applying ARM to the complete data set and to each of the sub-sets of the sample is that such approach allows for generating not only common rules that hold across the sample but also for identifying specific rules that hold only for each of the sub-sets of the sample.

### OPTION 2: ARM USING ONLY CONTEXTUAL VARIABLES

By utilizing this option, the investigator assesses the similarities and differences between the environments within which the entities under study operate. Just as in the case of the first option, we recommend running ARM twice – once using a complete data set, and then running the analysis using each of the discovered in Phase 1 sub-sets.

By applying ARM to the complete data set, we are likely to find the rules that hold in general for the whole sample. For example, if we are in the possession of such variables as "average household income", "average sizes of a household", or "average price of a house", then we should be able to generate the itemsets that help in separating the sample into sub-sets of "environmental peers" – the groups within the sample that operate within a similar context. Again, it is very useful to compare the results of ARM with the results of CA of Phase 1, for, while we would expect that the sub-sets of the sample operate in similar environments, it might turn out not being so. For example, if the sample was partitioned into "high-", "average-", and "low-"performing restaurants, then we have a reason to expect that, *ceteris paribus*, "high-"performing restaurants operate within a more advantageous socio-economic environment. However, it is also possible that "average" performers operate within "low" environment. If this is the case, then we have a reason to reassess whether "average" performer is indeed as "average" as the results of CA suggest.

By running ARM on each of the sub-sets separately, we would be able to generate itemsets that could be unique to each cluster. The results could be very useful in the case if the investigator wants to obtain a more precise picture of the environment for each group in the sample. For example, while running ARM using a complete set allows us to inquire into the differences in the environment between "high", "average", and "low" performers, it is by running ARM on each of the sub-sets we could see, more precisely, what their respective environments look like.

### Option 3: ARM Using Intrinsic and Contextual Variables

This option gives the researcher an opportunity of obtaining an insight into the impact of the context on the performance of the entity. While we expect that "high" performers are, indeed, superior to "low" performers in terms of the effectiveness and efficiency of operations, we also must acknowledge that if, hypothetically, we take "high" performers out of their context and place it in the context of "low" performers, then the situation may change. Simply put, context matters.

By running ARM using the complete data set, we hope to obtain some general insights into the interaction between the contextual variables and intrinsic variables and their impact on the level of performance (or whatever the factor is that resulted in *n*-cluster solution of Phase 1). While Option 1 allows us to connect the internal characteristics of an entity and the level of entity's performance, and Option 2 deals with the environment, this option allows us to consider internal and external factors together.

As in previous options for ARM in this phase, we recommend running ARM using the complete data set, followed by using the sub-sets of the sample.

## EXAMPLES OF APPLICATION OF CA AND ARM

**Chapter in this book:** 24.

**Some published papers:**

Samoilenko, S. and Osei-Bryson, K.M. (2019). Representation Matters: An Exploration of the Socio-Economic Impacts of ICT-Enabled Public Value in the Context of Sub-Saharan Economies, *International Journal of Information Management*, *49*, 69–85.

Samoilenko, S. and Osei-Bryson, K.M. (2020). Start a Business, Get a Credit, Make an Impact: Do ICTs Help? Impact of ICT on Legitimization of SMEs, *International Journal of Information Communication Technologies and Human Development*, *12*(2), 29–47.

# 12 C3: Heterogeneous Sample (Groupings Are Not Known) – CA, DTI, and MR

One of the pre-requisites for applying this module is the presence of a causal model that the researcher is interested in testing. However, the application of MR alone may not be the best choice of action, because of the possible heterogeneity of the sample that is manifested via two possible factors. First, there is a possibility of a threshold value in one or more variables of the MR model. Second, the presence of a contextual variable that may have an indirect impact on the variables comprising the MR model. In general, there are two possible types of data sets that the investigator may have at their disposal:

- First, the data set is limited to the number of the variables that comprise the MR model.
- Second, the data set that includes additional variables to those comprising the MR model.

Let us consider an illustrative scenario that could be used in application of this module. Suppose, we want to inquire into the factors impacting a semester's GPA score of students who work part-time. The MR model we are considering is:

*(# of courses enrolled, # of hours studied per week, # of hours worked per week)→GPA.*

In considering the relationship, we may suspect that there are some possible thresholds that we may encounter. For example, we would expect that the *# of hours* studied cannot increase continuously – at some point the hours of study do not positively impact the *GPA* of a student. Same is the case with *# of hours* worked per week, where it may not make a big difference to the score of *GPA* if a student works 5 or 6 or 7 hours per week, but we would expect a negative impact of the part-time work if the *# of hours* exceeds, let us say, 20.

Also, given the availability of contextual variables, we may consider, for example, whether or not number of hours students spend exercising or duration of the commute to work may play a role in impacting the relationship.

Beyond that, it is reasonable to suspect the difference in the relationships described by the MR model in the case of the students of different classifications. Given the same levels of the independent variables, their impact on GPA would differ depending on whether the subject is a freshman or a senior. Simply put, even if the causal model makes general sense, that is, "it works", a variety of factors may impact its specifics, that is, "how it works" and "when it works".

## PHASE 1: CA

Regardless of whether the data set is limited to the variables comprising the MR model or not, the process performing clustering is essentially the same. One thing to consider, of course, is whether to run *k*-means type of clustering or whether to select a hierarchical option. This decision, as well as the decision regarding the number of the resultant clusters in the case of *k*-means option, resides under the purview of the investigator.

Additionally, in the case of the data set comprising independent and dependent variables of the MR model and contextual variables, the investigator should decide whether to conduct CA using the complete data set, or whether to run CA using the data set limited to the variables of the MR model.

Also, given a large enough sample, it is also a good option to run three analyses – the first one using only the variables of the MR model, the second one using only contextual variables, and the third one using the complete data set comprised of the MR model and contextual variables.

## PHASE 2: DTI

Let us consider two options for Phase 2.

### Option 1: Data Set Is Limited to Variables of the MR Model

One of the decisions that needs to be made in the case of this option is whether to keep the dependent variable in the data set. If the DTI is run on a complete data set, then it is possible that one of the top splits may include the dependent variable of MR model. This may suggest the presence of a threshold value to the investigator, and if this is the case, then it could be of value to run MR using the sub-sets of the sample instead of the complete data set. However, if the data set is not very large, then it is not practical to do so. Consequently, the investigator may decide to remove the dependent variable from the data set to be subjected to DTI.

### Option 2: Data Set Comprises Variables of the MR Model and Contextual Variables

The important consideration to make if pursuing this option is whether to exclude MR variables from the data set. By excluding the independent and dependent variables, the investigator could identify those contextual variables that clearly differentiate the sub-groups. If this is the case, then a viable option would be to include the top-split contextual variables in the MR model. If this is done, then the investigator may consider running a regression analysis using the complete sample, as well as using each of the sub-groups separately.

If DTI analysis is run using the complete data set (e.g., MR and contextual variables together), then it may provide a valuable insight into the "hierarchy of the importance" of the top-split variables, for it may be that the top-level split is represented by a contextual variable, and the splits below are associated with the variables comprising the MR model.

## PHASE 3: MR

Given the original MR model, there are two decisions that the investigator may want to consider. The decisions are regarding the variables comprising the MR model and regarding the data sets to use for the analyses.

The first decision to make is whether or not to create an additional augmented MR model. This option is worth considering if one (or more) variable of the model was not among the top-split variables.

Also, if a top-level split has identified a contextual variable as being important in differentiating the clusters, then the investigator may consider adding this variable to the original MR model.

Let us consider that DTI yielded three top-level splits. One of the top-split variables is *duration of the commute to work* – a contextual variable, while other two are *# of hours studied per week* and *# of hours worked per week* – independent variables of the MR model. If this is the case, then it could be of benefit to test the original MR model, as well as two augmented ones, as follows:

1. *(# of courses enrolled, # of hours studied per week, # of hours worked per week) → GPA*
2. *(# of hours studied per week, # of hours worked per week) → GPA, and*
3. *(duration of the commute to work, # of hours studied per week, # of hours worked per week) → GPA.*

The second decision to consider is regarding the data set. Given the large enough data set, it could be of benefit to use multiple samples – a complete one comprising the clusters identified by CA, and the sub-sets representing each cluster.

It is also worth mentioning a consideration for the presence of complementarities in the model – a consideration of whether or not to include interaction terms in the original model. It is quite possible, that in the model # 3 *duration of the commute to work* and *# of hours worked per week* may end up being complementary, thus, it may be of interest to include the interaction term *duration of the commute to work * # of hours worked per week* in the model and test the significance of its impact on GPA.

For example, if the general model of MR takes form of:

$$Y = a + b1*X1 + b2*X2 + b3*X1X2 + \ldots + bn*Xn + bk*Xk + bm*Xn*Xk + \varepsilon,$$

then the test for interaction amounts to testing the null hypothesis H0: b3 = 0.

And in the case of b3 ≠ 0, we are able to reject the null hypothesis of no interaction between X1 and X2.

However, the interpretation of the interaction term in MR is not as straightforward as the interpretation of the slope coefficient of an independent variable (e.g., b1 or b2). For example, b3 in the equation above reflects the relationship between Y and X1 and X2 when X1 and X2 increase jointly. Furthermore, b3 in the equation above reflects conditional relationship between Y and X1 and X2, for the impact of X1 on Y would depend on the level of X2 and vice versa.

## EXAMPLE OF APPLICATION OF CA, DTI, AND MR

**Chapter in this book:** 18.
**Published paper:**

Samoilenko, S. and Osei-Bryson, K.M. (2013). Using Data Envelopment Analysis (DEA) for Monitoring Efficiency-Based Performance of Productivity-Driven Organizations: Design and Implementation of a Decision Support System. *OMEGA*, *41*(1), 131–142.

# 13 C4: Heterogeneous Sample (Groupings Are Not Known) – CA, DTI, and ARM

It is fair to say that this module is a good option to use, primarily, for exploratory purposes. The whole idea behind the itemsets generated by ARM is to find commonly occurring associations; consequently, this module does not allow for exploring any causal relationships, nor does it allow for generating testable hypotheses. However, this module is well-suited for generating insights allowing for describing naturally occurring sub-groups of the sample.

Let us consider an illustrative scenario for this module – the sample of interest is represented by student-athletes of Good University. Let us also consider two aspects of a university life of any student-athlete: academic and athletic. Some student-athletes get injured, others don't, and some student-athletes get in academic trouble (e.g., academic probation due to low GPA), while others do just fine academically. There are two perspectives that could be considered when discussing why student-athletes get injured or placed on probation. We can take a position that all injuries and academic problems are due to random events and there is no rhyme or reason for why the problems occur. Conversely, we may consider that while, indeed, there is an element of randomness present, there are also general similarities that are common to injured student-athletes and that the student-athletes on probation are not entirely unlike each other. If we select the second perspective when considering the issue, then we would like to discover the similarities via constructing some sort of a general profile of a student-athlete on probation or of those who are injured.

Consequently, it is reasonable to assume that the investigator (e.g., a representative of Student Success Center or an athletic trainer) would be interested in finding out what differentiates those students who get injured from those who don't, and those students who go on academic probation from those who do well in terms of their GPA. Such interest is well-justified, for injured student-athletes, as well as student-athletes on academic probation, are more likely to drop out of the university, and student withdrawal negatively impacts retention statistics and the stream of revenue that the students provide to the university. In the absence of causal models of a type "(x,y)→lowGPA" and "(a,b)→injury", all the investigator may rely on is a general description – a profile – of a student-athlete who is injured or on probation (we'll leave the case of "injured AND on probation" out).

Consequently, a purpose of using this module is to create a general descriptive picture of four types of student-athletes – those of good academic standing, on probation, injured, and healthy. Once the general characterizations have been created, the investigator would be able to assess the commonalities and differences between the risk groups, and, hopefully, propose some pre-emptive interventions for incoming and current student-athletes.

One of the requirements of this module is a data set comprised of a large number of contextual variables.

## PHASE 1: CA

The process of performing clustering is essentially the same as in previous modules. One thing to consider, of course, is whether to run *k*-means type of clustering or whether to select a hierarchical option. This decision, as well as the decision regarding the number of the resultant clusters in the case of *k*-means option, resides solely under the purview of the investigator. In the case of our scenario, we would like to arrive to a 4-cluster solution, so, *k*-means clustering with $k=4$ would be an option to consider. The question to answer is what variables should be removed from the data set prior to CA – if we have a data set that comprises a large number of contextual variables, then it is possible that the clustering will be based on variables that are not very helpful in creating an actionable profile of each group (e.g., height, weight, and years of experience and playing sport). Additionally, some important variables as zip-codes, classifications (e.g., sophomore, junior), and majors, while being significant to the later stages of the analysis, should be removed from the sample as well.

It is important to note that in this module, CA is not necessarily going to produce a solution that is congruent to the groupings of interest, such as, for example, Cluster 1 is "injured", Cluster 2 is "healthy", Cluster 3 is "good standing", and Cluster 4 is "probation". As a result, it is important to examine each cluster's membership in terms of the composition based on the intended groupings (e.g., injured, healthy, good standing, and probation). Thus, it could be beneficial to answer the following questions:

1. What is the composition of each cluster relative to the predefined categories?
2. Are there any similarities between clusters and categories in terms of the common variables?

Consequently, an outcome of Phase 1 is an *n*-cluster solution with a corresponding statistic, per cluster, regarding its membership in terms of each of the type of interest to the investigation.

## PHASE 2: DTI

We arrive at Phase 2 with two types of target variables to be created for DTI – let us discuss them in order.

### Option 1: A Priori Target Variable

The first type is based on generated a priori sub-sets of the sample. In the case of the illustrative example, a target variable "group" may be encoded as "1" is "healthy", "2" is "injured", "3" is "good standing", and "4" is "probation". Once this is done, the analysis can proceed. It is important to run the DTI using a complete data set, for it is possible that the top-level splits, when examined, produce some contextual commonalities for each node. It is reasonable to suppose that, for example, "injured" student-athletes may have in common "morning training sessions" and "morning classes", while those "on probation" may have similar "majors" and "number of games started". The outcome of the first option is a set of the top-split variables yielded by DTI, and the consequent assessment of each node in terms of the common contextual variables.

### Option 2: CA-based Target Variable

The second type of the target variable is based on the results of CA of the first phase, and the purpose of the second option is to identify the variables responsible for differentiating naturally (as in contrast to set a priori) occurring sub-groups in the sample. Consequently, if the results of Phase 1 yielded a 4-cluster solution, then "target" variable would have values of "one", "two", "three", and "four". As in the case of the first option, it is important to investigate similarities of the members of each node in terms of the contextual variables. It is likely, for example, that the sub-groups may differ based on such contextual variables as "type of sport", where we would expect that the number of injured football players would be much greater than the number of injured players in golf. The outcome of the second option is a set of top-split variables and the assessment of each naturally generated cluster in terms of the common contextual variables.

Overall, the outcome of Phase 2 is an assessment, of a compare-and-contrast type, of the membership of each type of the clusters in terms of the contextual variables. Specifically, it is important to look for an overlap between a priori and naturally occurring clusters in terms of the contextual variables describing them.

## PHASE 3: ARM

The goal of this phase is to discover a variety of itemsets that describe each of the sub-sets of the sample, under a desired condition that the itemsets could be compared vis-à-vis each other. Ideally, we would prefer the generated itemsets that are comparable in terms of a level, or a category of actionable variables.

For example, in the case of our scenario, we would like to discover itemsets of the type:

"freshman, football, on-campus→injured"
"junior, golf, commuter→healthy"

or

"early morning training, early morning classes, low incoming GPA→probation"
"early morning training, mid-day classes, mid incoming GPA→good standing".

If such itemsets are generated, then we could use them to construct profile corresponding to each cluster, and, consequently, use the profile not only for reactive (e.g., request a change in a schedule of a student) but also for predictive purposes (e.g., screen incoming and current students).

The last phase of this module may involve multiple ARM analyses, and we suggest performing them as a sequence of steps.

### Step 1

We suggest starting with the complete sample – this would allow for (possibly) generating itemsets that could be of general interest because they identify the associations that describe all student-athletes in the sample and those who are not group-specific.

### Step 2

The requirement for this step involves introducing a variable, let us say, "Cluster" that encodes the sub-groups identified by CA in Phase 1. Running ARM, in this case, may result in discovering the itemsets that could be compared across the clusters. Using this option, in our opinion, is important because it allows us to describe naturally-occurring groups of entities. We also suggest, in this case, running ARM twice – first, using the complete data set, and, second, using the sub-set of variables identified in Step 1.

### Step 3

Prior to the analysis, we need to create a variable, let us say, "Group", which allows for encoding the identified a priori categories of interest (e.g., "healthy", "injured", "good standing", and "probation"). Similar to Step 2, the analysis could be started with the complete sample without excluding any of the variables. After that we suggest running ARM using only the variables identified in Step 1.

### Step 4

The purpose of this step is to run the analysis that incorporates, within the data set, variables representing naturally occurring sub-groups of the sample (e.g., "Cluster") AND identified a priori categories of interest (e.g., "Group"). It could be of benefit to consider running ARM using four data sets:

1. Complete sample
2. Sub-set of the sample comprising the variables identified in Step 2
3. Sub-set of the sample comprising the variables identified in Step 3
4. Sub-set of the sample comprising the variables identified in Step 2 and Step 3.

## EXAMPLES OF APPLICATION OF CA, DTI, AND ARM

**Chapter in this book**: 23, 24.

**Some published papers:**

Samoilenko, S. and Osei-Bryson, K.M. (2019). Representation Matters: An Exploration of the Socio-economic Impacts of ICT-Enabled Public Value in the Context of Sub-Saharan Economies, *International Journal of Information Management*, *49*, 69–85.

Samoilenko, S. and Osei-Bryson, K.M. (2020). Start a Business, Get a Credit, Make an Impact: Do ICTs Help? Impact of ICT on Legitimization of SMEs, *International Journal of Information Communication Technologies and Human Development*, *12*(2), 29–47.

# 14 C5: Heterogeneous Sample (Groupings Are Not Known) – CA and DEA

One of the obvious pre-requisites for using this module is the presence of the defined data envelopment analysis (DEA) model, and a possible scenario that involves the situation where a decision maker wants to test the assumption of homogeneity of the sample to be subjected to DEA. It is worth mentioning that a domain area's knowledge by the investigator is indispensable for using this module successfully, for CA, being an unguided technique, would not offer any explanations for why the sub-groups exist (if they do) in the sample.

Let us consider a simple scenario for this module: we are interested in investigating relative efficiencies of the academic departments of a university. The DEA model that we are planning to use is as follows:

*(# of Faculty, # Years of Experience, Average)→(# of Courses/Semester, # of Students/Course).*

## PHASE 1: CA

The importance of CA to this module is clear: from the perspective of DEA, all DMUs in the sample are alike. Consequently, as long as the investigator declared the "semantic similarity" of the decision making units (DMUs) (academic departments in our example), the analysis can be conducted. However, there are always differences between real-life entities, and CA allows for gaining insights into general dimensions of those differences.

Let us also assume two possible scenarios regarding the data set to be subjected to CA, where in one case, the sample is limited to the variables comprising the DEA model, and in the second case, the data set comprises the variables of the DEA model as well as some contextual variables.

### OPTION 1: CA BASED ON THE DEA MODEL

In the case where the data set is limited to the variables of the DEA model, the analysis is straightforward, and a better option, perhaps, is to run a hierarchical analysis. This is because the DEA assumes the relative homogeneity of the sample, and if the investigator suspects the presence of heterogeneity a priori (e.g., via specifying the value of $k$ in $k$-means option), then a good reason should be presented for why the single DEA is conducted using the whole sample instead of multiple DEAs using the sub-sets of the sample.

In any case, the results of CA should be investigated by the decision maker – and, again, there is no substitute for having a knowledge regarding the domain area – in finding some sort of an explanation for the presence of the sub-groups, or, making sense of the resultant clusters.

## Option 2: CA based on the DEA Model and Contextual Variables

If the available sample comprises the variables of the DEA model and contextual variables, then we suggest running CA three times:

1. Using only the variables of the DEA model
2. Using only contextual variables
3. Using a complete sample.

Similar to Option 1, we suggest using a hierarchical analysis. Once the analyses are completed, the investigator should assess, using their domain area's knowledge, the sources of heterogeneity in the three types of samples above.

Let us consider some of the reasons for why the sample may not be homogeneous.

First, if we ran CA based on the DEA model only, then, given our illustrative scenario, it is possible that the sources of heterogeneity may be associated with the number of students per course. This could be due to some of the departments teaching general education courses, or introductory courses, which usually contain a much greater number of students enrolled. Simply put, the heterogeneity reflects the differences in the levels of the variable(s) comprising the DEA model.

Second, the resultant $n$-cluster solution in the case of using only contextual variables may suggest the presence of the sources of heterogeneity that, otherwise, would not be easy to note. For example, it is possible that the departments would differ, significantly, because some of them have research labs, while others don't, and some departments are heavily research-oriented, while others concentrate on teaching. As a result, this could be a scenario of the relative homogeneity of the sample based on the levels of the variables of the DEA model, albeit exhibiting signs of heterogeneity based on the external to the DEA model's factors.

Finally, by running CA based on the set comprising the DEA model and contextual variables, we could identify general differences between the sub-groups in the sample. For example, we can expect that there are differences between the groups of the departments in terms of the variables of the DEA model, and we can also expect that there are differences in terms of the contextual variables. However, we would also expect that those differences are, somehow, related – those are general differences. For example, some departments may offer only undergraduate degrees, while other departments may be offering undergraduate and master's degrees, while other departments may be offering degrees up to PhD level. Simply put, it is possible that we discover that the differences in the level of the variables of the DEA model are associated, or linked, with the differences in the level(s) of the contextual variables.

Let us consider a more interesting Option 2, where we could end up with three clustering solutions, one per type of the sample that was used to run CA.

## PHASE 2: DEA

Prior to conducting DEA, the investigator must make a decision regarding the data set to be subjected to DEA. Basically, the decision is to be made whether to use physical or logical partitioning of the sample.

Physical partitioning implies using the results of CA to partition the original data set into sub-sets based on the results of Phase 1. This will present additional evidence regarding relative homogeneity of the DMUs in each sub-set, but this is only a good option if the investigator gains insights into the nature of heterogeneity and concludes, as a result, that the DMUs in the sample should not be, despite their semantic similarity, treated as a single group.

Logical partitioning, on the other hand, implies no physical separation of the original sample into the sub-sets. Instead, the investigator simply notes the cluster that each DMU belongs to, and, once the DEA is completed, groups all DMUs into their clusters.

Given the options of Phase 1, the investigator may end up with three groupings of the results of the DEA.

Again, three types of groupings do not imply running three different analyses – it simply means running it once and grouping the results differently (e.g., based on three results of Option 2 of Phase 1).

From this point on, running DEA is a straightforward affair – select the orientation of the model, select the assumption regarding return to scale, and run the analysis. Once the scores of the relative efficiency of the DMUs in the sample have been obtained, scores should be grouped based on the results of the logical partitioning, and, then, averaged.

This will allow for a variety of comparisons of the groupings based on the levels of the variables of the DEA, levels of contextual variables, as well as the interrelationship between the two.

## EXAMPLES OF APPLICATION OF CA AND DEA

**Chapters in this book:** 19, 25, 26.

**Some published papers:**

Samoilenko, S. and Osei-Bryson, K.M. (2007). Increasing the Discriminatory Power of DEA in the Presence of the Sample Heterogeneity with Cluster Analysis and Decision Trees, *Expert Systems with Applications*, *34*(2), 1568–1581.

Samoilenko, S. (2008). Contributing Factors to Information Technology Investment Utilization in Transition Economies: An Empirical Investigation, *Information Technology for Development*, *14*(1), 52–75.

Samoilenko, S. and Osei-Bryson, K.M. (2010). Determining Sources of Relative Inefficiency in Heterogeneous Samples: Methodology Using Cluster Analysis, DEA and Neural Networks, *European Journal of Operational Research*, *206*, 479–487.

# 15 C6: Heterogeneous Sample (Groupings Are Not Known) – CA, DEA, and ARM

This module could be viewed as an extension of the previous one, albeit with one important difference.

Let us recall that a data envelopment analysis (DEA) model is not a "true" production model – it is simply an input–output model that makes sense to the decision maker and that is defensible to their audience. Consequently, while a previous module allows us to compare averaged scores of relative efficiency across different sub-groups, it does not allow us to describe the decision making units (DMUs) in the sample in terms of anything else but the variables of the DEA model. This module allows the investigator to do exactly that and ARM is a tool allowing for obtaining naturally occurring associations that may describe each sub-group in the sample. Clearly, one of the data set's requirements for this module is to be comprised of the variables of the DEA model AND contextual variables.

Let us consider the following scenario for the module: the sample on interest is represented by the sophomore students of a University's School of Business, and the DEA model selected by the investigator is:

"(#of courses enrolled/semester, #of hours of study/week) $\rightarrow$ semester GPA".

The research problem deals with the identification of the risk factors responsible for decreasing retention rate among the sophomore students caused by low GPA.

## PHASE 1: CA

Based on the scenario for this module, the investigator may suspect the presence of heterogeneity of the sample, where the sources of heterogeneity could be of two types – contextual and endemic. Under such circumstances, it is worthy of consideration to run both *k*-means clustering and hierarchical clustering.

Let us consider the contextual factors first.

Any school of business comprises different departments. And, as a result, it is quite possible that sophomores of the Department of Economics are different from those from the Department of Marketing and the Department of Information Systems. On the other hand, it is also possible that there are many undecided sophomores, or, even if the sophomores have had declared their majors, they are taking only general education courses during their first two years in the school. In any case, the domain area knowledge is required to suspect the presence of the heterogeneity that may exist in the sample, where potential context-specific factors may suggest the presence of $k$ sub-groups of the sample.

If this is the case, then it is reasonable to select the value of $k$ = *# of the departments in the school*, or, in the case of "declared vs. undecided" majors to select the value of $k = 2$. It is worth mentioning again that the selected a priori value of $k$ should be, in some way or form, justified via the specificity of context. And, as long as this could be clearly articulated and supported by solid reasoning, the choice of the value of $k$ is not likely to be rejected by the readers.

On the other hand, the investigator may want to assume that the contextual differences suggested above are not very relevant; instead, it is more important to concentrate on the endemic differences that exist between the students. Let us consider some of the reasons for such differences. First, it could be that the differences are associated with freshman, or even high school GPA of the students. Second, it is possible that the differences are associated, hypothetically, with geographic areas where the students come from (e.g., one region/area has weaker schools than the other area). It is also possible that there could be differences between students staying on-campus vs. commuter students living off-campus. Finally, there are differences between student-athletes and non-athletes.

In the case of our interest in endemic factors impacting the heterogeneity of the sample, a hierarchical clustering would yield a solution that may provide some insights that would be lost in the case of $k$-means option.

Finally, there could be differences between the students in terms of only the variables of the DEA model. For example, we may suspect that the number of courses taken by the students per semester differs. Or, that it is a number of hours spent studying that is different. Additionally, we may suspect that the student population of interest could be sub-divided on the basis of the GPA. If this is the case, then it is useful to consider running CA using only the data set comprised of the variables of the DEA model.

Needless to say, the investigator may consider performing CA multiple times by running $k$-means and hierarchical options. By exercising this option, the researcher may be able to gain insights into contextual and endemic factors that contribute to heterogeneity of the sample.

## PHASE 2: DEA

Running the analysis in this phase is a straightforward affair, where the only decisions to be made are regarding the assumption of the return-to-scale and the orientation of the model. In the case of the scenario for this module, input-oriented DEA under VRS would be an appropriate choice of options.

Once the results are obtained, the investigator should group the scores according to the results of Phase 1. If CA yielded multiple solutions, where, for example, $k$-means resulted in two- and five-cluster solutions, and hierarchical clustering resulted in three-cluster solution, then the scores of the relative efficiencies obtained in this phase should be grouped in three different ways. The goal is, of course, is to compare and rank the averaged values of the scores for each cluster.

At this point, the investigator should be able to identify the relatively more, as well as relatively less, efficient sub-sets of the DMUs in the sample. One of the goals of Phase 2 is to match the results of the DEA to the results of CA – it could be very useful to identify those clustering solutions, which result in the greatest difference between the clusters in terms of the averaged scores of the DEA.

Simply put, we would like to identify a cluster containing DMUs with very high scores of relative efficiency, and a cluster with very low scores (and may be a cluster with "in-between" scores). If such solution is identified, then it is of benefit to investigate what criteria were responsible for such clustering solution.

## PHASE 3: ARM

The purpose of Phase 3 is to generate two types of itemsets.

First, we would like to obtain the general rules that hold for the whole sample. For example, in the case of the scenario for this module, we would like to discover itemsets of the type:

> "(LOW_of courses enrolled/semester, HIGH_of hours of study/week) → HIGH_semester GPA",
> "(HIGH_#of courses enrolled/semester, LOW_#of hours of study/week) → LOW_semester GPA".

Simply put, such associations offer actionable information in terms of the variables of the DEA model.

However, we would also like to generate the rules that include the contextual variables that may have some relationship with the DEA model. Consequently, the two levers that the investigator has in Phase 2 are a choice of sample and the variables included in or excluded from the sample. This results, overall, in four options.

### Option 1: Complete Sample, # of Variables = the DEA Model

The first option would allow for generating itemsets that are common to the sample and are based on the variables of the DEA model. The results may allow for finding associations within the DEA model that actually exist, such as "LOW_# of hours of study/week → LOW_semester GPA" or "HIGH_# of courses enrolled/semester → HIGH_of hours of study/week". Simply put, by using this option, we aim to generate level-based (in terms of the values of the DEA model) itemsets.

### Option 2: Complete Sample, # of Variables = the DEA Model + Contextual Variables

The second option allows for finding associations between contextual variables and the variables of the DEA model. For example, if using our running example, it could be of benefit to generate itemsets of the types:

> "(High_Contextual_Var1, Low_Contextual_Var2) → LOW_#of hours of study/week"

or

> "(Low_Contextual_Var3, Low_Contextual_Var4) → HIGH_semester GPA".

### Option 3: Sub-Sets of the Sample, # of Variables = the DEA Model

Given *n*-clusters generated during Phase 1, the investigator performs an *n* number of analyses, where the main goal is to identify itemsets that characterize each of *n* clusters. It is possible that the itemsets would differentiate, clearly, each of *n*-clusters in terms of the levels of the variables of the DEA model.

However, it also possible that no such differentiation would be generated. If this is the case, then the investigator may consider going back to Phase 1 in order to re-evaluate the number of the generated clusters (e.g., to consider a two-, instead of a four-, cluster solution).

### Option 4: Sub-sets of the Sample, # of Variables = the DEA Model + Contextual Variables

By relying on this option, the investigator is bound to identify itemsets that differentiate *n*-clusters in terms of the contextual variables or the variables of the DEA model. After all, Phase 1's *n*-cluster solution is an indicator of the level-based heterogeneity of the sample.

For example, in the case of the running scenario for this module, it could be of benefit to generate itemsets of the types:

For Cluster 1: "(High_Contextual_Var1, LOW_#of hours of study/week) → LOW_semester GPA",
"(High_Contextual_Var1, Low_Contextual_Var2) → LOW_#of hours of study/week",

and

For Cluster 2: "(Low_Contextual_Var3, HIGH_#of hours of study/week) → HIGH_semester GPA",
"(High_Contextual_Var2, Low_Contextual_Var4) → HIGH_#of hours of study/week".

As our reader can see, the main purpose of this module is to describe the sub-groups in the sample that differ in terms of the relative efficiency of conversion of inputs into outputs. And, in the case if we are interested in some sort of intervention directed at the improvement of the lower performing sub-group, an additional challenge is to describe each of the clusters in terms of actionable variables.

## EXAMPLES OF APPLICATION OF CA, DEA, AND ARM

**Chapters in this book**: 23, 24.

**Some published papers:**

Samoilenko, S. and Osei-Bryson, K.M. (2019). Representation Matters: An Exploration of the Socio-Economic Impacts of ICT-Enabled Public Value in the Context of Sub-Saharan Economies, *International Journal of Information Management*, *49*, 69–85.

Samoilenko, S. and Osei-Bryson, K.M. (2020). Start a Business, Get a Credit, Make an Impact: Do ICTs Help? Impact of ICT on Legitimization of SMEs, *International Journal of Information Communication Technologies and Human Development*, *12*(2), 29–47.

# 16 C7: Heterogeneous Sample (Groupings Are Not Known) – CA, DTI, and DEA

The purpose of this module, in simple terms, is to investigate whether a possible presence of the sample heterogeneity has an impact on the differences in the levels of relative efficiencies of the decision making units (DMUs) in the sample. Let us consider an illustrative scenario for this module where we want to investigate the level of performance of different retail locations of the same type. Let us further suppose that we are using the following data envelopment analysis (DEA) model:

"(#_of_Employees/day, #_of_Deliveries/day) → #_of_Purchases/day".

For all intents and purposes, based on the *Input → Output* model, we know that the "high performers" are those locations that have the smallest number of daily employees and inventory deliveries (given input orientation of the model). However, we may also suspect that a geographic area of a retail location may play a role in addition to climate and socio-economic characteristic of the surrounding area.

Consequently, we may want to conduct the inquiry in the sequence of the following steps:

1. Test the assumption of homogeneity of the sample (CA).
2. Identify the factors responsible for the heterogeneity of the sample (DTI).
3. Assess the relationships between the differentiating factors and the variables of the DEA model.
4. Compare the levels of the relative performance of the groups comprising the sample (DEA).

## PHASE 1: CA

Overall, the purpose of applying CA in this module is to discover naturally occurring groupings based on, first, the differences in the levels of the variables of the DEA model, and, second, on the differences in the levels of contextual variables. Additionally, we would like to test the overall assumption of homogeneity of the sample – based on the data set comprising all available data.

Consequently, given that the investigator is in possession of the sample consisting of the variables of the DEA model and contextual variables, it could be of benefit to consider running CA three times – first, using only the variables of the DEA model, second, using only contextual variables, and, third, using a complete sample.

The justification for such suggestion is intuitive – the three options would (possibly) result in three different solutions, where each solution is based (again, possibly) on different factors. This may not be important during the first phase of the inquiry, but it will become important during Phase 2.

Another decision to be made is regarding the type of the clustering to perform – *k*-means vs. hierarchical. Just like in previous modules, if *k*-means option is selected, then the investigator should have a justification ready for why the given value of *k* is selected.

Once the analyses have been performed, the researcher should compare and contrast the resultant solutions, with the purpose of identifying common factors and associations that could be responsible for heterogeneity of the sample. While this may be a fairly straightforward exercise in the case of the first two options, the results of CA based on the complete sample could produce some interesting results, where, for example, the clusters would differ in terms of a contextual variable "Population_Density", which could have an association with the variable of the DEA model "#_of_Employees/day". After all, we would expect that less densely populated geographic areas would have less retail workers than the densely populated areas would.

## PHASE 2: DTI

Prior to conducting DTI, the investigator must create a target variable for each of the three solutions produced by CA of Phase 1. The domain of values, of course, would be dependent on the number of clusters produced by each solution. The outcome of the second phase is *n* sets of top-split variables, given that three cluster analyses resulted in *n* solutions each.

The important step in this phase is to assess the relationships between the top-split variables and the variables of the DEA model. In the case when CA was performed on the data set comprised only of the variables of the DEA model, which resulted in *n*-cluster solution, the top-level splits would be based on those variables. However, in the case of the sample comprised of the context variables the splits would be based on other variables, and the domain knowledge of the investigator is required to make a possible connection between the variables of the DEA model and the top-split variables.

## PHASE 3: DEA

Prior to conducting DEA, the investigator must make decisions regarding the orientation of the model and the preferred assumption regarding returns to scale. After that, the investigator may want to perform a logical partitioning of the sample, where DEA is conducted using the complete set of the data (e.g., input and output variables of the DEA model), but each DMU is "assigned" to its own cluster based on the results of Phase 1. Once the scores of the relative efficiency have been calculated, the results are grouped according to the results of CA solutions yielded by Phased 1.

For example, if Phase 1 yielded three different solutions based on the composition of the sample (e.g., variables of the DEA model, contextual variables, variables of the DEA model, AND contextual variables), where each solution resulted in three clusters, then the investigator would end up with three sets of the averages per solution ($3 \times 3 = 9$ overall). The goal here is, of course, to identify the clustering solution that yields the greatest difference between the clusters in terms of the averaged scores of the relative efficiency of the DMUs in each cluster.

## EXAMPLES OF APPLICATION OF CA, DTI, AND DEA

**Chapters in this book:** 18, 25, 26.

**Some published papers:**

Samoilenko, S. and Osei-Bryson, K.M. (2007). Increasing the Discriminatory Power of DEA in the Presence of the Sample Heterogeneity with Cluster Analysis and Decision Trees, *Expert Systems with Applications*, *34*(2), 1568–1581.

Samoilenko, S. (2008). Contributing Factors to Information Technology Investment Utilization in Transition Economies: An Empirical Investigation, *Information Technology for Development*, *14*(1), 52–75.

Samoilenko, S. and Osei-Bryson, K.M. (2013). Using Data Envelopment Analysis (DEA) for Monitoring Efficiency-Based Performance of Productivity-Driven Organizations: Design and Implementation of a Decision Support System. *OMEGA*, *41*(1), 131–142.

# 17 C8: Heterogeneous Sample (Groupings Not Known) – CA, DTI, DEA, and NN

Similar to data envelopment analysis (DEA)-based modules previously covered in this text, this combination of methods allows for calculating the scores of the relative efficiency of the decision making units (DMUs) in the sample, as well as for comparing the averaged scores across groups of the DMUs. Unlike any other combination of methods, however, this module allows for discerning the sources that give rise to the differences in averaged scores. A large data sample, however, is a necessary requirement for utilizing this module successfully. To illustrate the applicability of this module, we offer the following illustrative example:

Let us consider two groups of basketball players – HighScorers and LowScorers. Both groups are described by the same DEA model "(#MinutesPlayed, #ShotsAttempted) → #PointsScored". However, HighScorers differ from LowScorers in terms of the levels of inputs and outputs; it turned out that HighScorers get more points than LowScorers do, but they also play more minutes and attempt more shots than LowScorers. Meaning, there are level-based differences between two groups in terms of the variables of the DEA model. Let us further suppose that the results of the DEA yielded that HighScorers, as a group, are relatively more efficient than LowScorers.

Within such situation, the difference in the averaged scores of the relative efficiency between two groups could be due to two possible reasons. First, it could be that HighScorers have a higher level of transformative capacity than LowScorers – they are, indeed, more efficient in converting number of minutes played and attempted shots into the number of points (simply put, they are a group of *better* players). However, it is also possible that the difference is due to the differences in the level of inputs (LowScorers are a group of *less played* players), and, if LowScorers would be given an opportunity to play more minutes and take more shots, then they would become as efficient as HighScorers and would score more points.

Finding the reason for the differences becomes important if we are interested in improving the level of relative efficiency of LowScorers, because we have two options to consider – first, we could let them play more minutes and encourage them to take more shots, hoping that it would result in corresponding increase in points. Or, we could concentrate on making LowScorers better players, and once they improve, we could raise the levels of their inputs by allowing them to play more minutes and encouraging them to take more shots.

So, as a running scenario for this module, let us consider a sample of sales offices owned by Company X and described by the following DEA model:

"(#ofEmployees, #ofAllocatedAccounts) → #ofPlacedOrders)".

We would like to analyze performance of the offices vis-à-vis each other as well to get some insights into the ways of improving performance of the offices and, as a result, of Company X.

## PHASE 1: CA

The purpose of CA is to discover naturally occurring groups in the sample – we would like to find out whether or not sales offices belong to distinct groups that differ from each other. We assume that the investigator is in possession of a data set comprised of the variables of the DEA model and contextual variables.

Given the presence of the domain knowledge by the investigator, *k*-means clustering is an appropriate method to proceed if there is a reason to suspect the presence of *k* distinct groups of sales offices. Otherwise, a hierarchical clustering method could also be considered. Regarding the composition of the data set, it could be useful to run two CA – one using a complete data set, and the other using a sub-set comprising only the variables of the DEA model.

The benefit of doing so is in identifying contextual variables that may have an impact on applicability and appropriateness of the DEA model. For example, let us suppose that we ended up with 2-cluster solutions in both cases (e.g., complete data set and the DEA-only data set). If the membership of the clusters of "complete data set" is very similar (clearly, the criterion of "similarity" must be established and defended by the investigator) to the clusters of the "DEA model only" data set, then we have a reason to think that the environment within which the offices operate is pretty much homogeneous. However, if the memberships would turn out to be very different, then we have a reason to suspect that the environment may need to be considered when formulating and applying the DEA model.

Let us suppose that we examined two clusters of the complete data set and determined that they differ, and significantly so, in terms of the values of the variables "population density" and "average income". Meaning, the membership of the first cluster is characterized by lower levels of population density and income then the membership of the second one. It is quite plausible that the levels of these two contextual variables have an association with the input variables of the DEA model, where offices in the less populated areas characterized by lower levels of income would also have less number of employees in the office and have a smaller number of allocated accounts, and vice versa. It is not clear if there is a generic way of considering such situation (e.g., via implementing a weighting scheme); instead, a domain matter knowledge of the investigator is required to take into consideration a possible impact of the contextual variables on the variables of the DEA model.

Barred the homogeneity of the sample, let us assume that the outcome of Phase 1 is 2-cluster solution, which means that there are two distinct groups of sales offices of Company X. If the results of Phase 1 did not yield a multi-cluster solution, then the investigator should consider a suitability of this module for her purpose.

## PHASE 2: DTI

The main goal of DTI in this module is to determine whether the analysis yields top splits, based on the variables of the DEA model, that clearly differentiate multiple clusters. There are some points to consider – one is an acceptable level of purity of the node. It is important to decide what ratio would be considered acceptable for a given node. The second point is regarding the membership of a node – what would be a minimum overall membership of a node, and what would be acceptable number for a majority and minority membership of a node. These questions must be explicitly dealt with by the investigator and once the decision is made, the justification for a chosen set of criteria should be provided.

Let us suppose that the outcome of Phase 2 is a set of top-split variables that differentiate the sub-groups of the sample in terms of the levels of the variables of the DEA model; as a result, our 2-cluster solution is represented by "low-level" and "high-level" clusters.

## PHASE 3: DEA

Prior to conducting the analysis, the investigator must decide on the orientation of the model and the assumption regarding the return-to-scale, and the justification for the selected choices should be presented to a reader.

Another decision to make prior to conducting DEA is to whether to perform physical or logical partitioning of the sample. We advocate for using both options – let us consider some of the benefits for doing so.

Given the presence of two clusters that differ in terms of the levels of the DEA model' variables, the investigator could select to run DEA for each cluster. In such case, the results of DEA yield relatively efficient and relatively inefficient DMUs for "high-level" and "low-level" clusters. Let us also suppose that the results indicated that the averaged score of the relative efficiency of "high-level" cluster was higher than that of the "low-level" cluster.

What could we do with this information? One of the possible ways of using it is for the purposes of benchmarking. If we want to improve the level of performance of relatively inefficient DMUs of "low-level" cluster, then in order to do so, it could be of benefit for such DMUs to benchmark performance of the relatively efficient DMUs of the same cluster, rather than to attempt to emulate relatively efficient performers of "high-level" cluster.

In the case of the logical partitioning of the sample, the investigator performs DEA using a complete data set, where each of the DMUs is noted as being a member of its cluster. Once the scores are obtained, the averaged scores for each cluster are calculated and then compared. Let us suppose the results yielded higher averaged score of relative efficiency for "high-level" cluster and a lower averaged score for "low-level" cluster. The question arises: what factor is responsible for the difference in the scores? Do we have a discrepancy because of the superior level of transformative capacity? Or, is it due to the difference in the levels of the variables of the DEA model? The answer to this question is important, because if we want to improve the level of performance of a DMU from "low-level" cluster relative to the level of performance of a DMU from "high-level" cluster, then we face two possible options. First, we could allocate resources toward increasing the level of the variables of input–output model – for example, we could increase the number of "Employees" and "Allocated Accounts" for "low-level" offices and hope that this will result in the increase of "Placed Orders". Or, we could allocate resources toward the development of the productivity of the workforce of "low-level" offices – we could spend money on training programs, better equipment, and so on. The next phase allows us to identify the preferred course of actions.

## PHASE 4: NN

In our discussion of this phase, we leave out decisions regarding such NN-specific issues as the selection of the algorithm and partitioning of the sample into sub-sets for the purposes of training, testing, and validation. Instead, we only discuss the issues relevant to the purpose of this module. The structure of NN model for this phase is based on the structure of the DEA model, where input variables of the DEA model become input nodes of NN, and the output variables of the DEA model become output nodes of NN.

Conceptually, this phase involves four steps – we describe them in order.

### STEP 1: CREATING AN NN MODEL OF "LOW-LEVEL" CLUSTER

The purpose of this step is to create a model of transformation of the inputs (represented by the inputs of the DEA model) into the outputs (outputs of the DEA model) of the members of "low-level" cluster. The output of the first step is a *saved* NN model for "low-level" cluster – the hidden nodes of the NN model.

### STEP 2: CREATING AN NN MODEL OF "HIGH-LEVEL" CLUSTER

The purpose of the second step is to create the NN model for the members of "high-level" cluster, and the mechanism is similar to that of Step 1. The output of the second step is a *saved* NN model for "high-level" cluster – the hidden nodes of the NN model.

Consequently, the outcome of the first two steps of Phase 4 is two NN models depicting transformation of inputs into outputs for two clusters.

### STEP 3: SIMULATION OF THE OUTPUTS OF "LOW-LEVEL" CLUSTER USING NN MODEL OF "HIGH-LEVEL" CLUSTER

The purpose of Step 3 is to generate outputs of "low-level" cluster based on its own level of inputs, but using NN model of transformation of inputs into outputs of the "high-level" cluster that was generated in Step 2. This, for all intents and purposes, allows the investigator to determine what the outputs of "low-level" cluster would

As a result, the investigator would be in the possession of evidence according to which the level of performance of "low-level" cluster could be improved vis-à-vis the level of performance of "high-level" cluster.

## EXAMPLES OF APPLICATION OF CA, DTI, DEA, AND NN

**Chapters in this book:** 18, 19, 25, 26.

**Some published papers:**

Samoilenko, S. and Osei-Bryson, K.M. (2007). Increasing the Discriminatory Power of DEA in the Presence of the Sample Heterogeneity with Cluster Analysis and Decision Trees, *Expert Systems with Applications*, *34*(2), 1568–1581.

Samoilenko, S. (2008). Contributing Factors to Information Technology Investment Utilization in Transition Economies: An Empirical Investigation, *Information Technology for Development*, *14*(1), 52–75.

Samoilenko, S. and Osei-Bryson, K.M. (2013). Using Data Envelopment Analysis (DEA) for Monitoring Efficiency-Based Performance of Productivity-Driven Organizations: Design and Implementation of a Decision Support System. *OMEGA*, *41*(1), 131–142.

Samoilenko, S. and Osei-Bryson, K.M. (2010). Determining Sources of Relative Inefficiency in Heterogeneous Samples: Methodology Using Cluster Analysis, DEA and Neural Networks, *European Journal of Operational Research*, *206*, 479–487.

be if its members were as efficient (utilized their model of transformation) as t members of "high-level" cluster.

Meaning, the investigator would be able to determine what would be the numb of "Placed Orders" for the members of "low-level" cluster be if they were to kee their own numbers of "Employees" and "Allocated Accounts", but were as efficie as the members of "high-level" cluster". The original input nodes of NN (inputs c the DEA model) and the simulated values of the output nodes of NN model need t be saved and will become a part of a new data set.

### Step 4: Simulation of the Outputs of "High-Level" Cluster Using NN Model of "Low-Level" Cluster

This step is a mirror image of Step 3, and the purpose of it to determine what would the level of outputs of "low-level" cluster be if its members were able to use higher levels of inputs without changing anything else (e.g., keeping their own model of transformation). Consequently, the investigator applies the values of the input nodes of the members of "high-level" cluster to the NN model of transformation of the members of "low-level" cluster generated in Step 1. Thus, as a result of this step, the investigator would be able to find out what the outputs of the members of the "low-level" cluster would be if its members used the inputs of the members of "high-level" cluster, but used their own level of efficiency of conversion of inputs into outputs. The values of the input nodes of NN (inputs of the DEA model) and the simulated values of the output nodes of NN model need to be saved and will become a part of a new data set.

## PHASE 5: DEA

The investigator arrives to this stage with the same DEA model, plus two new data sets for the lower performing cluster. If using our running example for this module, then the researcher has additional two data sets for "low-level" cluster that represent two scenarios, where outputs are based on:

1. The same levels of inputs applied to a superior level of transformative capacity, and
2. Higher levels of inputs applied to the original (native) level of transformative capacity.

Once data sets are ready, the investigator should re-run Phase 3 in order to determine in which case the averaged level of relative efficiency of "low-level" cluster improved the most.

For all intents and purposes, the investigator would end up with two new averaged scores of relative efficiency for the lower performing cluster. Once the scores are compared vis-à-vis each other, the determination could be made whether the reason for the lower scores of the relative efficiency is due to the lower levels of the inputs, or whether it is due to a lower level of transformative capacity.

# Section III

## Methodological Modules – Examples of Their Application

# 18 A Hybrid DEA/DM-based DSS for Productivity-Driven Environments

## INTRODUCTION

Modern organizations typically operate in dynamic, competitive environments. Within this context, the critical issues of organizational survival and advancement often lead to calls for improvements in the levels of effectiveness and efficiency of performance. However, due to the relativity of the concepts of efficiency and effectiveness, productivity-driven organizations must take into consideration the performance of their competitors. This requirement is due to the dynamic nature of the business environment which will cause the levels of performance of competing organizations to change over time, and if the efficiency of the competitors has improved, then a productivity-driven organization must respond with its own improvements in efficiency.

A desired capability of an organization to successfully respond to efficiency-related challenges suggests the need, first, for an effective mechanism that allows for discovering appropriate productivity models for improving overall organizational performance and, second for a feedback-type mechanism that allows for evaluating multiple productivity models in order to select the most suitable one.

The dynamic nature of the business environment also suggests the presence of a concept that is central to a productivity-driven organization, namely, that of the *superior stable configuration.* Given the goal of achieving a high level of efficiency of conversion of inputs into outputs, a superior stable configuration in the context of a productivity-driven organization may imply *a model of conversion of inputs into output (input–output model) characterized by a high level of efficiency.*

Overall, a decision maker tasked with a responsibility of improving performance of productivity-driven organization existing within a dynamic business environment must take into consideration internal (organizational) and external (environmental) factors. Similarly, if a decision making is to be added by an Information System, then the designers of such system must implement two sets of functionalities: *externally oriented* and *internally oriented.* The *externally oriented* functionality is directed toward evaluating the external competitive environment of a productivity-driven organization, as well as identifying the differences between the current state of the organization and the states of its competitors. The *internally oriented* functionality, on the other hand, is directed toward the optimization of the level of productivity of the organization, as well as toward an identification of the factors impacting the efficiency of the input–output process.

In this chapter, we will describe a decision support system (DSS) that allows assessing and managing the relative performance of organizations. Specifically, we focus on organizations that consider the states of their internal and external organizational environment in the formulation of their strategies, such that the achievement of an organizational goal is dependent on the level of performance that is commonly measured in terms of the levels of the efficiency of utilization of inputs, effectiveness of the production of outputs, and efficiency of conversion of inputs into outputs.

## DESCRIPTION OF THE DSS

The focus on the efficiency assessment suggests that an important component technique of the DSS is data envelopment analysis (DEA). However, other techniques are also required for providing answers to several questions that are relevant to the organization's search for the productivity model that is most suitable with respect to survival and advancement. Next, we outline how a DSS could be implemented using a combination of parametric and non-parametric data analytic and data mining techniques including DEA, cluster analysis (CA), decision tree (DT), neural networks (NN), and multivariate regression (MR).

Let us discuss how each of the above mentioned methods, alone, or in combination with other methods, could be used in the DSS. First, we will discuss externally oriented functionality.

### EXTERNALLY ORIENTED FUNCTIONALITY

CA allows for segmentation of the data set into naturally occurring heterogeneous groups. An application of this method allows for detecting the presence of multiple disparate groups of competitors in the external business environment. A decision maker can also determine whether the clusters comprising the data set differ in terms of the relative efficiency of utilization of inputs or production of outputs – DEA will help in this regard. By specifying a DEA model and running the analysis, we can obtain scores of the relative efficiency for each cluster, as well as see how the scores differ between the clusters.

If the data for the same group of competitors available for two points in time, let us say, *Year 1* and *Year 2*, then a decision maker can obtain insights regarding possible changes in the number of clusters, as well as changes in the membership of the clusters. This will allow for determining whether the structure of the competitive environment has changed over time. If DEA follows CA, then a decision maker can also identify the changes that took place in regard to relative efficiencies of the members comprising the data set.

After conducting CA and DEA, a decision maker may want to inquire into the existence of the factors that are possibly responsible for the presence of multiple groups of competitors, namely, what are some of the reasons for the heterogeneity of the external business environment? DTs can offer some help in this regard; once CA helped to identify the presence of the various groups comprising the sample, a target variable (let us say, *Cluster#*) will allow for identifying every member of the data set in terms of its membership in a given group. Then we can run DT analysis specifying that target variable – this will allow for determining the dimensions in the data set that differentiate the clusters the most. By comparing the results of DT analysis of the data set representing two points in time, we can determine, based on changes in differentiating dimensions, the possible reasons for the changes not only in the number of clusters, but also in the composition of the clusters.

By creating a new target variable reflecting the differences in the scores of the relative efficiency of each cluster (e.g., *LessEfficient* vs. *MoreEfficient*), we can also use the same DT-based approach to identify the differences between the clusters in regard to the scores of the relative efficiency. By conducting the same analysis at multiple points in time, we can identify the changes in regard to the relative efficiency, as well as factors possibly associated with identified changes.

At this point, we have obtained insights regarding the nature of the competitive environment – we identified the naturally occurring groupings, as well as changes in the groupings over time. We determined some of the variables that are responsible for separating the clusters, as well as the variables that were responsible for change in groupings over time. We also found out important information regarding the disparity between the clusters in terms of the relative efficiency, as well as some of the variables that are possibly responsible for the disparity.

Let us summarize the capabilities afforded by various combinations of CA, DEA, and DT to a decision maker for the purposes of assessing the external environment of an organization.

When the data analysis is conducted at *one point in time*:

- A combination of CA and DT allows for identification of naturally occurring groups of competitors, as well as determining major dimensions that differentiate the groups.
- A combination of CA and DEA allows for comparison of the naturally occurring groups in terms of the relative efficiency of conversion of inputs into outputs.
- A combination of CA, DEA, and DT allows for identification of factors differentiating relatively less efficient and relatively more efficient groups of competitors.

Additionally, when the data analysis is conducted at *multiple points in time*:

- A combination of CA and DT allows for identification of changes that took place in regard to the composition of naturally occurring groups of competitors, as well as for determination of the factors associated with the changes.
- A combination of CA and DEA allows for identification of the changes that took place in regard to the relative efficiency of the naturally occurring groups.
- A combination of CA, DEA, and DT allows for identification of factors associated with the changes in relative efficiency of the group that took place over time.

### Internally Oriented Functionality

Once we completed the analysis of the external environment of the organization, we need to turn our attention to the analysis of its internal state. All organizations are similar in the way that every one of them converts resources into products or services – this allows a decision maker to model an organization as a set of inputs (e.g., resources) and a set of outputs (e.g., products and/or services). MR analysis allows us to identify those independent variables (inputs) and their interaction terms (complementarities) that produce statistically significant impact on the dependent variable (output). It is, of course, a limitation of MR that the analysis allows for a single output variable. However, if a decision maker wants to consider multiple outputs, then this requirement could be accomplished by using multiple MR models.

Additionally, we can use NNs to create a model of input–output transformation of the organization. This requires identifying a set of inputs and outputs – the two sets will become, respectively, input and output nodes of NN. Once NN is trained, we can identify an *input–output transformation* model specific to the organization. By saving this model and varying levels of inputs we can determine the impact of the manipulation of inputs on the level of outputs. More importantly, we can use NN for the purposes of benchmarking – this will require creating a NN model of a better performing peer. In doing so, we can apply existing levels of the inputs of the organization to the transformation model of the better performing organizations to discern whether the improvements in performance could be obtained via the variation of the levels of inputs, or via improvements in the input–output transformation model.

Moreover, once we obtained via NN a set of simulated inputs and outputs, we can subject the set to DEA. In this situation a decision maker has two options in regard to input and output variables. First, the same set of inputs and outputs model could be used for a DEA model and NN analysis. Second, the variables used in DEA model could represent a subset of inputs and outputs used in NN analysis. In either case the simulated via NN set of inputs and outputs could be used to create a representation of two *simulated organizations*. The first simulation represents a level of outputs based on the manipulated inputs, but with the original input–output transformation process. The second set represents a set of outputs based on the original set of inputs, and manipulated input–output transformation process that was adapted from a better performing organization. Adding the simulated data to the original data set and then running DEA will allow for determining the impact of the simulation on the relative efficiency of utilization of inputs and production of outputs. As a result, a decision maker will be able to determine whether the improvements in efficiency should be obtained via variation in the levels of inputs, or whether the improvements should come from the changes in input–output transformation process.

At this point we can summarize the capabilities afforded by utilization of MR, DEA, and NN to a decision maker in regard to the analysis of the internal state of an organization.

When the data analysis is conducted at *one point in time*:

- MR allows for determining resources that have a significant impact on the output of the business process, as well as for determining the presence of synergies and complementarities of the resources on the output.

- NN allows for determining whether the level of the output of the business process should be increased by means of increasing the level of inputs, or by means of improving the process of converting resources into outputs.
- Combination of NN and DEA allows for determining whether the improvements in the relative efficiency of the organization should come from changes in the level of consumed resources or from the changes in the business process by which resources are converted into outputs – products or services.

Also, when the data analysis is conducted at *multiple points in time*:

- MR provides insights into the possible changes in regard to significance of the impact of resources, as well as their complementarities, on the outputs of the business processes.
- NN offers indications whether the process of input–output transformation of the organization has changed over a period of time.
- DEA and NN provide evidence regarding the changes that took place in regard to the relative efficiency of the organization, as well as offer insights into whether the identified changes are due to the changes of the levels of inputs and outputs, or due the changes in the organizational input–output transformation process.

## Architecture of the DSS

At this point we can combine the sequences of the data analytic methods that were outlined above within a comprehensive design of a single DSS (see Figure 18.1).

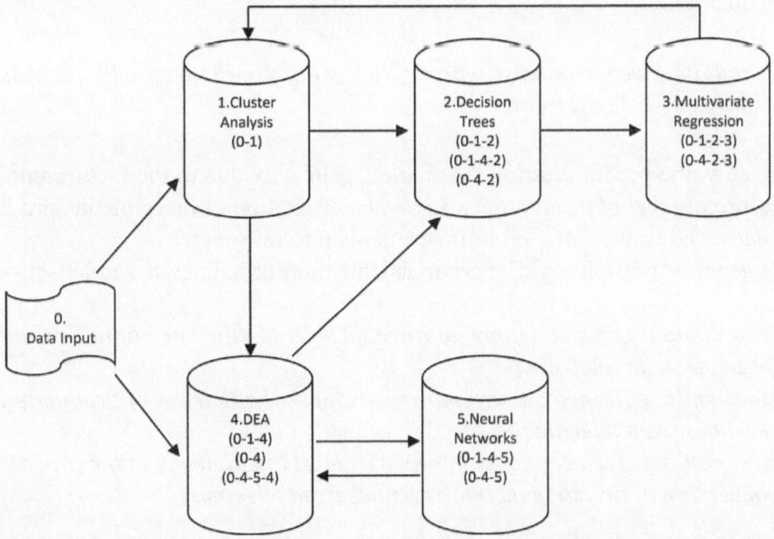


**FIGURE 18.1** Sequential method utilization within the design of the hybrid DEA/DM-based decision support system.

## AN ILLUSTRATIVE APPLICATION

The outlined DSS could be used in the context of any set of economic units, as long as the units are represented via the common DEA model. However, the analytical engine of the DSS should not be limited to a data set that is comprised only by inputs and outputs of the DEA model – a decision maker can obtain a richer set of insights if other variables are also included in the data set. As a consequence of the reliance of the DSS on the common DEA model, the system could be applied at the different levels of granularity – the DSS could be used at the departmental, organizational, industry, and country levels of analysis.

This illustrative example involves the country level and deals with the efficiency and effectiveness of the impact of investments in telecoms, a type of investments that is common to almost all of the economies in the world. The context is represented by the following 18 economies: Albania, Armenia, Azerbaijan, Belarus, Bulgaria, Czech Republic, Estonia, Hungary, Kazakhstan, Kyrgyz Republic, Latvia, Lithuania, Moldova, Poland, Romania, Slovakia, Slovenia, and Ukraine. The time-series data covering the period from 1993 to 2002 were obtained from the *World Development Indicators* database and the International Telecommunication Union' *Yearbook of Statistics.*

Within the context of the sample, 18 economies are economic entities characterized by the same business process – that of conversion of investments in telecoms into revenues. Because economies compete for Foreign Direct Investments and private investments, they are forced to compete with each other based on their levels of productive efficiency – conversion of investments into revenues. Clearly, the more productive economy would attract a larger pool of investment resources than the less productive one. Hence, a decision maker may pose the following general question:

*How could a given economy improve its level of productivity with regards to its investments in telecoms?*

Undoubtedly, the posed question is complex, primarily due to the factors impacting the measure of *level of productivity – namely*, utilization of investments, production of revenues, and transformation of investments into revenues.

Thus general question could be expanded into three efficiency-based sub-questions:

1. *How could a given economy improve its level of efficiency of utilization of investments in telecoms?*
2. *How could a given economy improve its level of efficiency of production of revenues from telecoms?*
3. *How could a given economy improve its level of efficiency of the process of conversion of investments into revenues from telecoms?*

We use SAS' *Enterprise Miner* data mining software to conduct CA, DT, NN, and MR, and *OnFront* to conduct DEA. Because the design of the proposed DSS systems is DEA-centric, one of the prerequisites for using it is associated with identifying a DEA model that is to be used in evaluating productivity of the organizational entities in the sample. We list the set of variables used in the illustrative example below.

***Input variables:***

- GDP per capita (in current US $)
- Full-time telecommunication staff (% of total labor force)
- Annual telecom investment per telecom worker
- Annual telecom investment (% of GDP in current US $)
- Annual telecom investment per capita
- Annual telecom investment per worker

***Output variables:***

- Total telecom services revenue per telecom worker
- Total telecom services revenue (% of GDP in current US $)
- Total telecom services revenue per worker
- Total telecom services revenue per capita

Next, we demonstrate, in a step-by-step fashion, an application of the DSS to the context of our illustrative example.

## Step 1: Is the Business Environment Homogeneous?

The purpose of the first step is to offer the decision maker a capability to inquire into the nature of the competitive business environment in regard to the presence of the multiple heterogeneous groups of competing business entities. As the reader may recall we implement this functionality by incorporating CA into the design of our DSS (Figure 18.2).

We began the CA by using "Automatic" setting, which did not require a specification of the exact number of clusters by the analyst. This setting produced a five-cluster solution that was considered to be the starting point in the analysis.

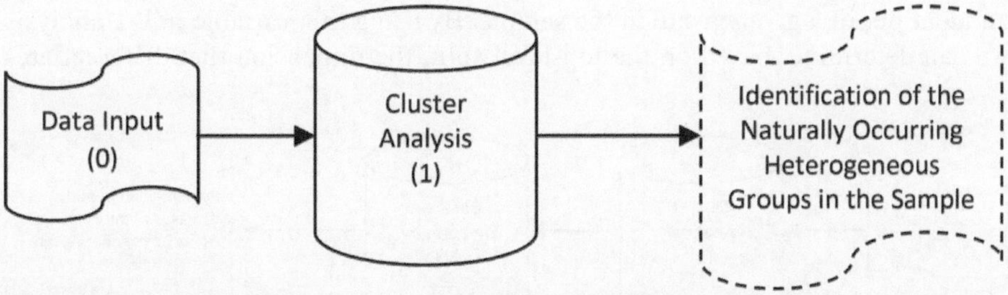


**FIGURE 18.2** Detection of changes in the external competitive environment.

By sequentially reducing the number of clusters we derived a two-cluster solution which was considered to be final; the result is provided below.

**Cluster1:** Czech Republic, Estonia, Hungary, Latvia, Lithuania, Poland, Slovenia, Slovakia
**Cluster2:** Albania, Armenia, Azerbaijan, Belarus, Bulgaria, Kazakhstan, Kyrgyzstan, Moldova, Romania, Ukraine

However, what is the basis for accepting this two-cluster segmentation versus some other grouping? We suggest using an external evaluation approach to assess cluster validity, where a domain expert opinion can provide external confirmation of the validity of this segmentation. In the case of the illustrative example such domain expert support is provided by Piatkowski (2003), who concluded that in the period "between 1995 and 2000 ICT capital has most potently contributed to output growth in the Czech Republic, Hungary, Poland, and Slovenia". Thus, it could be suggested that we were able to separate 18 economies into the two groups: the *Leaders* group (Cluster1) that consists of economies which benefited the most from the investments in telecom, and the *Followers* group (Cluster2) that consists of economies where the benefits are less pronounced.

Furthermore, the results of the CA offered evidence that Cluster1 is different from Cluster2 in terms of the two dimensions: *Investments* and *Revenues from telecoms.* Consequently, with regard to these dimensions, for a given economy its own cluster will represent a peer context, while members of the other cluster will comprise a non-peer context. Overall, Step 1 allowed the decision maker to determine that the competitive business environment comprised of 18 economies is not homogenous, but appears to be comprised of two groups that differ in terms of investments and revenues from telecoms.

## Step 2: What Are the Factors Responsible for Heterogeneity of the Business Environment?

However, even if the decision maker identified the presence of multiple groups comprising a given business environment, it is not clear what differentiates the peer context from the non-peer context; this question will be answered by the functionality of our DSS described in Step 2 (Figure 18.3).

The results of the CA allow us to introduce a target variable *Cluster#* to serve as an identifier of a given group in the sample. By using this variable in DT analysis we can determine, based on the top-level split, the dimension that differentiates

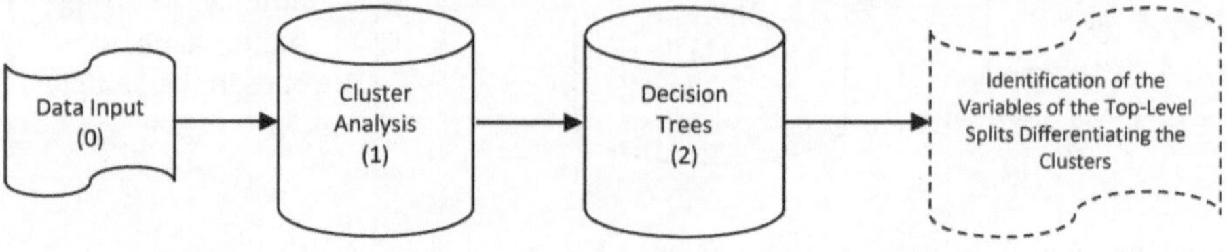


**FIGURE 18.3** Identification of the factors that differentiate groups of competitors.

the *Leaders* from the *Followers*. Clearly, when conducting DT analysis, we don't have to be limited to the set of variables that was used for CA. The results of DT analysis in the form of the decision rules are presented below.

```
IF Annual telecom investment     Cluster = {2: 96.4%; 1: 3.6%}
(Current US $per telecom         where N = 110.
worker) < $9,610 THEN

IF Annual telecom investment     Cluster = {2: 2.9%; 1: 97.1%}
(Current US $per telecom         where N = 70.
worker) ≥ $9,610 THEN
```

These results allow the decision maker to identify the relevant dimension differentiates the peer from the non-peer context the most. This means that while the two clusters differ in terms investments and revenues from telecoms, the single most important dimension that differentiates two clusters is associated with the level of investments in telecoms per telecom worker. Step 2 allows the decision maker to determine that the two groups of 18 economies differ most significantly in terms of the respective levels of investments in telecoms per telecom worker.

## Step 3: Do Groups of Competitors Differ in Terms of the Relative Efficiency?

Once the decision maker identified the presence of heterogeneous groups of the competitors within the business environment it is reasonable to inquire whether the groups differ in terms of the efficiency of utilization of investment and production of revenues. This additional information is obtained by means of incorporating DEA in the design of the proposed DSS (Figure 18.4).

Completing Step 3 involves running DEA and calculating the scores of the relative efficiency for each entity in the sample. It should be noted that DEA is not applied separately to the *Followers* and the *Leaders*. Instead, DEA is applied to

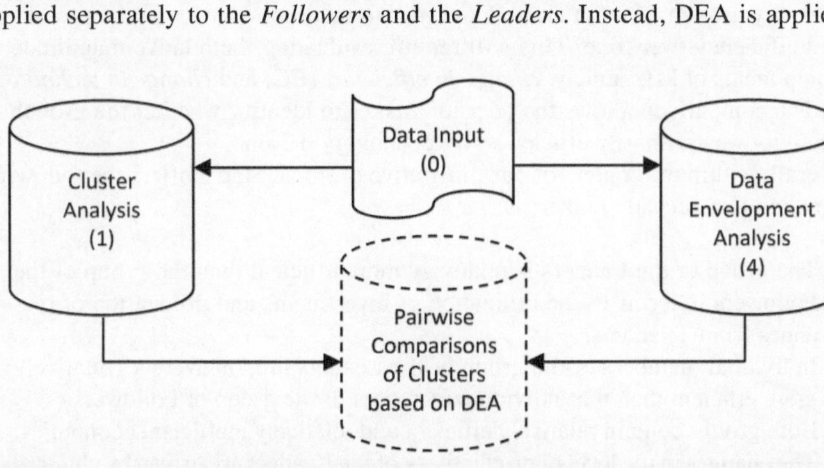


**FIGURE 18.4** What are the differences in relative efficiency among peer and non-peer groups?

**TABLE 18.1**
**Assessment of the Differences Between the Clusters in Terms of the Relative Efficiency**

| Orientation | Return to Scale | Cluster1 | Cluster2 | Conclusion |
|---|---|---|---|---|
| Input-oriented | CRS | 0.89 | 0.79 | *Cluster1* is relatively more efficient than *Cluster2* |
| | VRS | 0.95 | 0.88 | *Cluster1* is relatively more efficient than *Cluster2* |
| | NIRS | 0.89 | 0.80 | *Cluster1* is relatively more efficient than *Cluster2* |
| Output-oriented | CRS | 1.21 | 1.44 | *Cluster1* is relatively more efficient than *Cluster2* |
| | VRS | 1.18 | 1.30 | *Cluster1* is relatively more efficient than *Cluster2* |
| | NIRS | 1.21 | 1.38 | *Cluster1* is relatively more efficient than *Cluster2* |

the entire sample, and so the relative efficiency scores are not determined based on a cluster membership. Our application of DEA resulted in the scores of the *relative efficiency* for each entity in the entire set. If the scores are averaged per cluster, then the decision maker has the information regarding the averaged relative efficiency for the peer versus non-peer groups. For the purposes of our illustrative example, we conducted DEA under assumptions of *constant* (CRS), *variable* (VRS) and *non-increasing* (NIRS) *return-to-scale* and averaged the scores for the *Leaders* (Clusterl) and the *Followers* (Cluster2). The results are presented in Table 18.1.

During Step 3 we can also conduct DEA to calculate the *Malmquist index* (MI) of productivity growth for both clusters in order to measure changes in the productivity and efficiency over time. This will require evaluating the relative magnitude of the components of MI, namely, *change in efficiency* (EC) and *change in technology* (TC). The comparison allows the decision maker to identify whether the growth in productivity was primarily efficiency, or technology-driven.

Overall, within the context of our illustrative example Step 3 offers the following information to a decision maker:

- The group of the Leaders is relatively more efficient than the group of the Followers in terms of the utilization of investments and production of revenues from Telecoms.
- Individual members of the group of the Leaders are, on average, relatively more efficient than the Individual members of the group of Followers.
- Both groups contain relatively efficient and relatively inefficient economies.
- The changes in the level of productivity of the Leaders are driven by changes in efficiency, while the changes in the level of productivity of the Followers are driven by changes in technology.

As the reader can see, the results of Step 3 provide the decision maker with important information regarding the relative efficiency of the peer vs. the non-peer group within the competitive business environment. In the case of our illustrative example an investigator can easily determine that under any assumption of return to scale the *Leaders* are relatively more efficient than the *Followers* in terms of, both, utilization of investments and the production of revenues. However, we can expect that each cluster will contain relatively efficient economies and relatively inefficient ones. The purpose of the next step is to provide the decision maker with the functionality allowing inquiring into the differences between relatively inefficient and relatively efficient peer- and non-peer economies.

## Step 4: What Are some of the Factors Associated with the Differences in Relative Efficiency?

Because in the case of our illustrative example we ended up with two clusters (the *Leaders* and the *Followers*), we can identify four groups of economies within our sample. The groups are: relatively efficient economies of the *Leaders*, relatively inefficient economies of the *Leaders*, relatively efficient economies of the *Followers*, and relatively inefficient economies of the *Followers*. By introducing a target variable *ClusterEfficiency*, with domain of values [1, 2, 3, 4], we can identify each group of economies within each cluster and use the target variable in DT analysis to identify the split variables and their values that differentiate the groups. To identify the most meaningful splits the decision maker can opt to display the resulting DT in the form of the easy-to-interpret decision rules, and then to concentrate on the rules that have a high probability for the occurrence of the group based on the decision rule. Figure 18.5 illustrates Step 4 and the results are presented in Table 18.2.

The results demonstrate that the functionality provided by the DSS allows the decision maker to obtain important information regarding some of the factors that differentiate not only efficient and inefficient peers, but also efficient and inefficient non-peers. This information could be useful for the purposes of intra-group benchmarking, as well as for the purpose of formulating strategies for business units that are interested in intergroup transitioning.

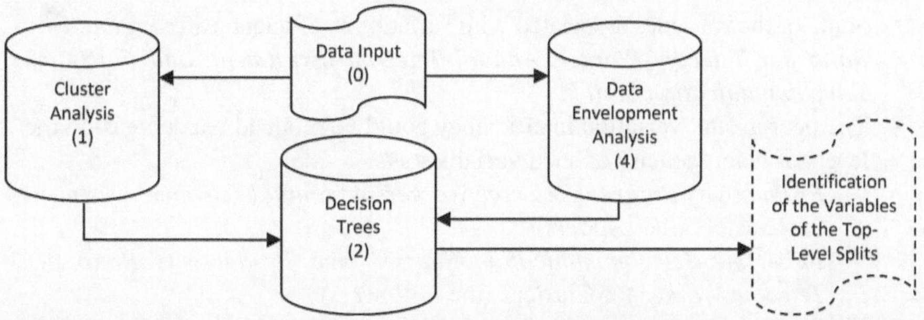


**FIGURE 18.5** What are some of the factors associated with the differences in relative efficiency?

**TABLE 18.2**
**Decision Rules Generated by Decision Tree Analysis**

| Group | Decision Rule | Posterior Prob. |
|---|---|---|
| Group 1: Efficient Leaders | *Productivity Ratio per Telecom Worker* ≥ 1.5674014075 & *Annual Telecom Investment* < $836,899,003 & *Full-Time Telecommunication Staff %* ≥ 0.0039016912 & *Annual Telecom Investment per Worker* ≥ $58 | 0.94 |
| | *Productivity Ratio per Telecom Worker* ≥ 4.1754445351 & *Annual Telecom Investment per Worker* ≥ $58 | 1.00 |
| Group 2: Inefficient Leaders | *Annual Telecom Investment* ≥ $836,899,003 & *Full-Time Telecommunication Staff %* ≥ 0.0039016912 & Productivity *Ratio per Telecom Worker* < 4.1754445351 & *Annual Telecom Investment per Worker* ≥ $58 | 1.00 |
| | *Full-Time Telecommunication Staff %* < 0.0039016912 & *Productivity Ratio per Telecom Worker* < 4.1754445351 & *Annual Telecom Investment per Worker* ≥ $58 | 1.00 |
| Group 3: Efficient Followers | *Full-Time Telecommunication Staff %* < 0.0031414015 & *Productivity Ratio per Telecom Worker* ≥ 3.8043909395 & *GDP per Capita* ≥ $519 & *Annual Telecom Investment per Worker* < $33 | 1.00 |
| | *Total Telecom Services Revenue* ≥ 0.0118204323 & GDP *per Capita* < $519 & *Annual Telecom Investment per Worker* < $33 | 1.00 |
| Group 4: Inefficient Followers | *Productivity Ratio per Telecom Worker* < 3.8043909395 & *GDP per Capita* ≥ $519 & *Annual Telecom Investment per Worker* < $33 | 1.00 |
| | *Productivity Ratio per Telecom Worker* < 2.002357802 & $33 ≤ *Annual Telecom Investment per Worker* < $58 | 1.00 |

Taken together, results of Step 4 allow the decision maker to obtain the following information regarding the heterogeneity of the sample in terms of the efficiency-based performance:

- Some of the variables associated with variation in efficiency are: *Productivity Ratio per Telecom Worker, Annual Telecom Investment, and Full-Time Telecommunication Staff %.*
- The peer group' variation in efficiency could be reduced via decreasing the levels of heterogeneity of such variables as:
  - *Productivity Ratio per Telecom Worker* and *Annual Telecom Investment* for Clusterl (the *Leaders*)
  - *Full-Time Telecommunication Staff %* and *Productivity Ratio per Telecom Worker* for Cluster2 (the *Followers*)
- The non-peer' group variation in efficiency could be reduced via decreasing the levels of heterogeneity of such variables as: *Productivity Ratio per Telecom Worker* and *Full-Time Telecommunication Staff %.*

At this point the decision maker is aware of the variables that are associated with the differences in regard to the efficiency-based performance; however, an additional benefit could be obtained from identifying complementarities between those variables that produce a synergistic effect on the output. The functionality of the DSS outlined in Step 5 allows the decision maker to identify some of the complementarities that may exist between the relevant to the production process variables.

### Step 5: Are There any Complementarities Between the Relevant Variables?

In order to identify existing complementarities between the production process variables we construct a three-variable model consisting of two inputs – investments and labor, and one output – GDP (Figure 18.6).

This will allow us to relate investments in telecoms, full-time telecom staff, and GDP as the following production function:

$$\text{GDP} = f(\text{investments in telecoms, full-time telecom staff}), \text{represented as } Y = f(K, L).$$

This allows us to construct the following formulation to test for the presence of interaction:

$$\log Y == \beta_0 + \beta_1 * \log K + \beta_2 * \log L + \beta_3 * \log K^2 + \beta_4 * \text{Log} L^2 + \beta_5 * \log K * \log L + \xi,.$$

A test for the presence of the interaction between investments in telecoms and telecom staff would involve testing of the following hypothesis:

H0: β5 is not statistically discernible from 0 at the given level of α.

The testing of the null hypothesis yielded the following results:

In the case of the *Leaders* β estimate is 57.4954 (**p @ 95%** is less than 0.0001) – this allows us to reject the null hypothesis of no interaction, and

In the case of the *Followers* β estimate is -2.1280 (**p @ 95%** is 0.0087) – the null hypothesis cannot be rejected.

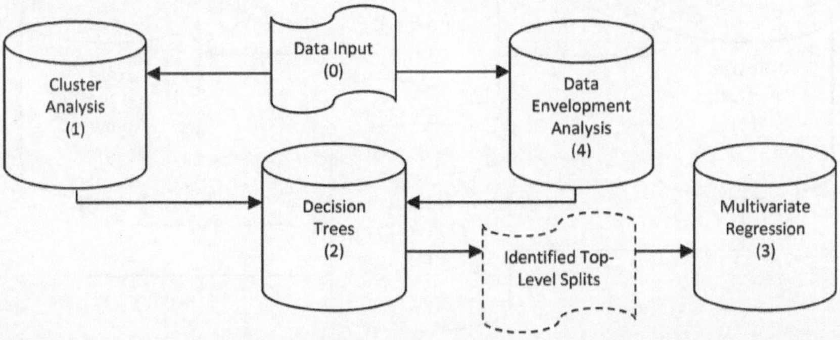


**FIGURE 18.6** What are some of the factors impacting the current levels of the relative efficiency?

Overall, Step 5 allows the decision maker to conclude the following:

> *Annual Telecom Investment and Full-Time Telecommunication Staff are complementary factors that may allow for improving the level of the efficiency-based performance of the organizational entities in the Leaders group (i.e. Cluster1).*

However, despite obtaining important insights regarding the presence of complementarities, the decision maker will still need additional information regarding the best route to improvements in the level of the efficiency of the production process, specifically as it relates to the production of outputs. For example, we determined that the *Leaders* are, on average, more efficient than the *Followers*; however, we also determined that the levels of investments and revenues of the *Leaders* are higher than those of the *Followers*.

This situation allows for two interpretations; first, members of the *Leaders* are more efficient than members of the *Followers* because of the superior process of conversion of inputs into outputs, or, the members of the *Leaders* are more efficient because they have higher levels of inputs which allows for establishing and maintaining more efficient processes. Consequently, the design of DSS must allow for the functionality allowing for determining the most appropriate route to improvement in the production of outputs, namely, whether to increase the level of inputs, or whether to improve the production processes first.

## STEP 6: WHAT IS A BETTER WAY TO IMPROVE PRODUCTION OF OUTPUTS?

Let us recall that NN analysis allows the decision maker to create a model of the input–output process in the form of an *input–output transformation* function, which then could be saved and applied to a new set of production inputs with the purpose of generating a new set of the production outputs (Figure 18.7).

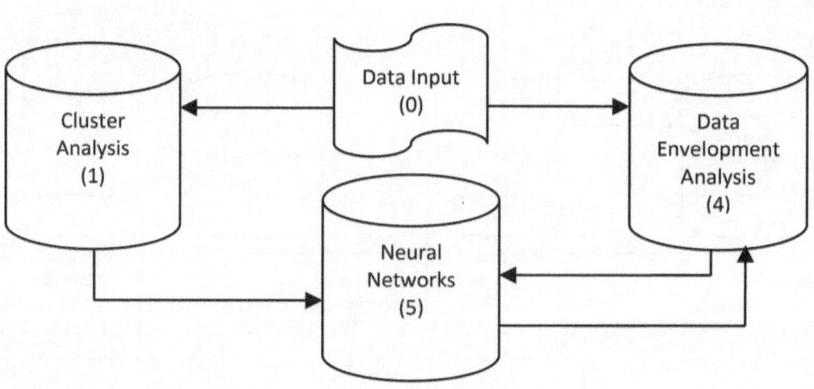


**FIGURE 18.7** What is the most effective way of increasing efficiency of the input–output process?

In the case of our illustrative example NN analysis allows us to generate two transformation functions, *TF1* for the *Leaders* and *TF2* for the *Followers*. If we apply the inputs of the *Leaders* to TF2 we can simulate the level of outputs that members of the *Followers* would have produced if they had the levels of inputs of the *Leaders*. Conversely, if we apply TF1 to inputs of the *Followers* we can simulate the level of outputs that members of the *Followers* would have obtained if they utilized process of conversion of inputs into outputs of the *Leaders*. Both simulated DEA models were created for the purposes of gaining insights into the most appropriate route of improving the level of efficiency of the *Followers*. Altogether, given 2 clusters, we end up with four DEA models:

| | |
|---|---|
| DEA Model 1 | Actual model based on the original inputs and outputs of the *Leaders* |
| DEA Model 2 | Actual model based on the original of inputs and outputs of the *Followers* |
| DEA Model 3 | Simulated model of the *Followers*, where the outputs are based on the inputs of the *Leaders* |
| DEA Model 4 | Simulated model of the *Followers*, where the outputs are based on the process of input–output conversion of the *Leaders* |

By comparing the scores produced by the original models with the scores of the simulated models we can determine whether the *Followers* would get a greater gain in efficiency of production of outputs from increasing the level of inputs (DEA Model 3), or from improving the efficiency of conversion of inputs into outputs (DEA Model 4). Results of this comparison are summarized as follows:

- An increase in the levels of inputs of the *Followers* to the levels of the *Leaders* results in a decrease in the scores of the relative efficiency of the *Followers*
- An improvement of the actual input–output transformation process of the *Followers* to that of the *Leaders* results in an increase in the scores of the relative efficiency of the *Followers*.

These findings allow a decision maker to determine that the economies of the *Followers*' group should not pursue an increase in the level of inputs as a mean of increasing efficiency of output production; instead, improvements in the production process should serve as a mean of increasing efficiency of production of output. Overall, the results of Step 6 offer evidence that:

*The existing inefficiencies of the Followers are associated with the inefficient processes of conversion of inputs into outputs, and not with the insufficient levels of inputs.*

## CONCLUSION

In this chapter, we presented a DEA-centric DSS that provides facilities for assessing and managing the relative performance of productivity-driven organizations that operate in unstable environments. The design of our DSS was guided by a set of requirements that are highly relevant to a productivity-driven organization's efforts to identify and evaluate multiple productivity models in order to select the most suitable one for the given organization. The resulting DSS is applicable to different organizational levels, including the country level and the firm level. In this chapter, we demonstrated the feasibility and usability of this DSS on country-level organizational entities.

## ACKNOWLEDGMENT

Material in this chapter previously appeared in: "Using Data Envelopment Analysis (DEA) for Monitoring Efficiency-Based Performance of Productivity-Driven Organizations: Design and Implementation of a Decision Support System", ***Omega 41:1***, 131–142 (2013).

## REFERENCES

Andre, F., Herrero, I., and Riesgo, L. (2010). A modified DEA model to estimate the importance of objectives with an application to agricultural economics, Omega, 38(5), 371–382.

Avkiran, N. and Rowlands, T. (2008). How to better identify the true managerial performance: State of the art using DEA, Omega, 36(2), 317–324.

Bezdek, J. (1981). Pattern Recognition with Fuzzy Objective Function Algorithms. Plenum Press, New York, NY.

Bishop, C.M. (1995). Neural Networks for Pattern Recognition. Oxford University Press, New York, NY.

Bock, H. (1996). Probabilistic models in partitional cluster analysis, Computational Statistics and Data Analysis, 23, 5–28.

Bollou, F. and Ngwenyama, O. (2008). Are ICT investments paying off in Africa? An analysis of total factor productivity in six West African countries from 1995 to 2002, Information Technology for Development, 14(4), 294–307.

Botti, L., Briec, W., and Cliquet, G. (2009). Plural forms versus franchise and company-owned systems: A DEA approach of hotel chain performance, Omega, 37(3), 566–578.

Bougnol, M.-L., Dula, J.H. Estellita Lins, M.P., and Moreira da Silva, A.C. (2010). Enhancing standard performance practices with DEA, Omega, 38(1–2), 33–45.

Bu, N., Fukuda O., and Tsuji, T. (2003). EMG-based motion discrimination using a novel recurrent neural network, Journal of Intelligent Information Systems, 21(2), 113–126.

Çelebi, D. and Bayraktar, D. (2008). An integrated neural network and data envelopment analysis for supplier evaluation under incomplete information, Expert Systems with Applications, 35(4), 1698–1710.

Chang, H., Choy, H., Cooper, W., and Ruefli, T. (2009). Using Malmquist Indexes to measure changes in the productivity and efficiency of US accounting firms before and after the Sarbanes-Oxley Act, Omega, 37(5), 951–960.

Charnes, A., Cooper, W.W., Lewin, A.Y., and Seiford, L.M. (1994). Data envelopment analysis: Theory, in Methodology and Applications. Kluwer Academic Publishers, Norwell, MA.

Choi, Y. S. and Yoo, S.I. (2001). Text database discovery on the web: Neural net based approach, Journal of Intelligent Information Systems, 16(1), 5–20.

Cook, W. and Bala, K. (2007). Performance measurement and classification data in DEA: Input-oriented model, Omega, 35(1), 39–52.

Cook, W. and Hababou, M. (2001). Sales performance measurement in bank branches, Omega, 29(4), 299–307.

Cook, W., Liang, L., and Zhu, J. (2010). Measuring performance of two-stage network structures by DEA: A review and future perspective, Omega, 38(6), 423–430.

Cook, W., Seiford, L., and Zhu, L. (2004). Models for performance benchmarking: Measuring the effect of e-business activities on banking performance, Omega, 32(4), 313–322.

Cook, W.D. and Zhu, J. (2008). Data Envelopment Analysis: Modeling Operational Processes and Measuring Productivity, ISBN/EAN13: 1434830233/9781434830234.

Cooper, W.W. and Tone, K. (1997). Measures of inefficiency in data envelopment analysis and stochastic frontier estimation, European Journal of Operational Research, 99(1), 72–88.

Cooper, W.W., Seiford, L.M. and Zhu, J. (2004). Data envelopment analysis: History, models and interpretations, in Handbook on Data Envelopment Analysis, Chapter 1, pp. 1–39, W.W. Cooper, L.M. Seiford and J. Zhu (eds.), Kluwer Academic Publishers, Boston, MA.

Crestani, F. and Rijsbergen, C. J. van (1997). A model for adaptive information retrieval, Journal of Intelligent Information Systems, 8(1), 29–56.

Cristofor, D. and Simovici, D. (2002). An information-theoretical approach to clustering categorical databases using genetic algorithms. In: Proceedings of the SIAM DM Workshop on Clustering High Dimensional Data, pp. 37–46. Arlington, VA.

Dave, R. (1992). Generalized fuzzy C-shells clustering and detection of circular and elliptic boundaries, Pattern Recognition, 25, 713–722.

Du, J., Liang, L., Chen, Y., and Bi, G. (2010). DEA-based production planning, Omega, 38(1–2), 105–112.

Dula, J.H. (2002). Data envelopment analysis (DEA), in: Handbook of Applied Optimization, P.M. Pardalos and M.G.C. Resende (eds.), pp. 531–543, Oxford University Press, New York, NY.

Eilat, H., Golany, B., and Shtub, A. (2008). R&D project evaluation: An integrated DEA and balanced scorecard approach, Omega, 36(5), 895–912.

Emrouznejad, A. and Shale E. (2009). A combined neural network and DEA for measuring efficiency of large scale datasets, Computers and Industrial Engineering, 56(1), 249–254.

Fu, L. (1999). Knowledge discovery by inductive neural networks, IEEE Transactions on Knowledge and Data Engineering, 11(6), 992–998.

Harb H. and Chen, L. (2005). Voice-based gender identification in multimedia applications, Journal of Intelligent Information Systems, 24(2–3), 179–198.

Hirschberg, J.G. and Lye, J.N. (2001). Clustering in a data envelopment analysis using bootstrapped efficiency scores. Department of Economics – Working Paper Series 800, The University of Melbourne.

Holsheimer, M. and Siebes, A.P.J.M. (1994). Data mining: The Search for Knowledge in Databases, Report CS-R9406, Centre for Mathematics and Computer Science, Amsterdam, The Netherlands.

Jain, A., Murty, M., and Flynn, P. (1999). Data clustering: A review. ACM Computing Surveys, 31(3), 264–323.

Kao, C. (2010). Malmquist productivity index based on common-weights DEA: The case of Taiwan forests after reorganization, Omega, 38(6), 484–491.

Kao, C. and Hung, H. (2009). Efficiency analysis of university departments: An empirical study, Omega, 36(4), 653–664.

Khouja, M. (1995). The use of data envelopment analysis for technology selection, Computers and Industrial Engineering, 28(1), 123–132.

Lemos, C.A.A., Lins, M.P.E., and Ebecken, N.F.F. (2005). DEA implementation and clustering analysis using the K-means algorithm, in Data Mining VI – Data Mining, Text Mining and their Business Applications, Transactions of the Wessex Institute.

Liu, J. and Lu, W. (2010). DEA and ranking with the network-based approach: A case of R&D performance. Omega, 38(6), 453–464.

Lozano-Vivas, A. and Pastor, J. (2010). Do performance and environmental conditions act as barriers for cross-border banking in Europe? Omega, 38(5), 275–282.
Lu, H., Setiono, R. and Liu, H. (1996). Effective data mining using neural networks, IEEE Transactions on Knowledge and Data Engineering, 8(6), 957–961.
McQueen, J. (1967). Some methods for classification and analysis of multivariate observations. In: Lecam, L.M. and Neyman, J. (eds.) Proceedings of the 5th Berkeley Symposium on Mathematical Statistics and Probability, pp. 281–297.
Morais P. and Camanho, A. (2011). Evaluation of performance of European cities with the aim to promote quality of life improvements, Omega, 39(4), 398–409.
Mostafa, M.M. (2009). A probabilistic neural network approach for modelling and classifying efficiency of GCC banks, International Journal of Business Performance Management, 11(3), 236–258.
Murtagh, F. (1983). A survey of recent advances in hierarchical clustering algorithms which use cluster centers, Computer Journal, 26, 354–359.
Okazaki, S. (2006). What do we know about mobile internet adopters? A cluster analysis, Information and Management, 43(2), 127–141.
Osei-Bryson, K.-M. and Inniss, T. (2007). A hybrid clustering algorithm, Computers & Operations Research, 34(11), 3255–3269.
Pao, Y.-H. and Sobajic, D.J. (1991). Neural networks and knowledge engineering, IEEE Transactions on Knowledge and Data Engineering, 3(2), 185–192.
Parthasarathy, S. and Anbazhagan, N. (2008). Evaluating ERP projects using DEA and regression analysis, International Journal of Business Information Systems, 3(2), 140–157.
Piatkowski, M. (2003). Does ICT Investment Matter for Output Growth and Labor Productivity in Transition Economies? TIGER Working Paper Series, No. 47. December, Warsaw. Retrieved January 20, 2008 from www.tiger.edu.pl.
Rai, A., Tang, X., Brown, P., and Keil, M. (2006). Assimilation patterns in the use of electronic procurement innovations: A cluster analysis, Information and Management, 43(3), 336–349.
Ramanathan, B. and Yunfeng, J. (2009). Incorporating cost and environmental factors in quality function deployment using data envelopment analysis, Omega, 37(3), 711–723.
Razi, M. and Athappilly, K. (2005). A comparative predictive analysis of neural networks, nonlinear regression and classification and regression tree (CART) models, Expert Systems with Applications, 29(1), 65–74.
Samoilenko, S. (2008a). Information systems fitness and risk in IS development: Insights and implications from chaos and complex systems theories, Information Systems Frontiers, 10(3), 281–292.
Samoilenko, S. (2008b). Contributing factors to information technology investment utilization in transition economies: An empirical investigation. Information Technology for Development, 14(1), 52–75.
Samoilenko, S. and Osei-Bryson, K.M. (2008). Increasing the discriminatory power of DEA in the presence of the sample heterogeneity with cluster analysis and decision trees, Expert Systems with Applications, 34 (2), 1568–1581.
Samoilenko, S. and Osei-Bryson, K.M (2008a). Strategies for telecoms to improve efficiency in the production of revenues: An empirical investigation in the context of transition economies, Journal of Global Information Technology Management, 11(4), 56–75.
Samoilenko, S. and Osei-Bryson, K.M. (2008b). An exploration of the effects of the interaction between ICT and labor force on economic growth in transitional economies, International Journal of Production Economics, 115(2), 471–481.
Samoilenko, S. and Green, L. (2008). Convergence and productive efficiency in the context of 18 transition economies: Empirical investigation using DEA. Proceedings of the Southern Association for Information Systems Conference, Richmond, VA, USA, March 13–15.

Seiford, L. and Zhu, J. (1999). An investigation of returns to scale in data envelopment analysis, Omega, 27(1), 1–11.

Shao B.B.M. and Lin W.T. (2001). Measuring the value of information technology in technical efficiency with stochastic production frontiers, Information and Software Technology, 43(7), 447–56.

Shimshak, D. Lenard, M., and Klimberg, R. (2009). Incorporating quality into data envelopment analysis of nursing home performance: A case study, Omega, 37(3), 672–685.

Shin, H.W. and Sohn S.Y. (2004). Multi-attribute scoring method for mobile telecommunication subscribers, Expert Systems with Applications, 26(3), 363–368.

Sohn, S. and Moon, T. (2004). Decision tree based on data envelopment analysis for effective technology commercialization, Expert Systems with Applications, 26(2), 279–284.

Stewart, T. (2010). Goal directed benchmarking for organizational efficiency, Omega, 38(6), 534–539.

Theodorou, P. and Florou, G. (2008). Manufacturing strategies and financial performance—The effect of advanced information technology: CAD/CAM systems, Omega, 36(1), 107–121.

Tone, K. and Tsutsui, M. (2010). Dynamic DEA: A slacks-based measure approach. Omega, 38(3–4), 145–156.

Vai, M. and Xu, Z. (1995). Representing knowledge by neural networks for qualitative analysis and reasoning, IEEE Transactions on Knowledge and Data Engineering, 7(5), 683–690.

Wallace, L. Keil, M. and Rai, A. (2004). Understanding software project risk: A cluster analysis, Information and Management, 42, 115–125.

Ward, J. (1963). Hierarchical grouping to optimize an objective function, Journal of the American Statistical Association. 58, 236–244.

Wiederhold, G. (1992). Mediators in the architecture of future information systems, IEEE Computer, 25(3), 8–49.

Wiederhold, G.C.M., Walker, M.G., Blum, R.L. and Downs S.M. (1986). Acquisition of knowledge from data. In: Proceedings of ACM SIGART ISMIS, pp. 74–84.

Wu, D. (2009). Supplier selection: A hybrid model using DEA, decision tree and neural network, Expert Systems with Applications, 36(5), 9105–9112.

Wu, M.-C., Lin, S.-Y. and Lin, C.-H. (2006). An effective application of decision tree to stock trading, Expert Systems with Applications, 31(2), 270–274.

Yu, M. and Lin, E. (2008). Efficiency and effectiveness in railway performance using a multi-activity network DEA model, Omega, 36(6), 1005–1017.

Zhong, W., Yuan, W., Li, S., and Huang, Z. (2011). The performance evaluation of regional R&D investments in China: An application of DEA based on the first official China economic census data, Omega, 39 (4), 447–455.

# 19 Determining Sources of Relative Inefficiency in Heterogeneous Samples

## *Methodology Using Cluster Analysis, DEA, and Neural Networks*

## INTRODUCTION

*Data envelopment analysis* (DEA) is a widely used non-parametric analytic tool (e.g., Chen and van Dalen, 2010; Khalili et al., 2010; Asmild et al., 2007; Shao and Lin, 2001; Gillen and Lall, 1997; Khouja, 1995; Doyle and Green, 1994) that is commonly applied in the research and practitioner communities to determine the relative efficiencies of the *decision making units* (DMU). Any entity that receives a set of inputs and produces a set of outputs could be designated as a DMU, thus, any group of such entities could be subjected to DEA. As a result, this method has been applied to evaluate productivity and performance of airports (Gillen and Lall, 1997; Martin and Roman, 2001; Pels et al., 2001), efficiency of US Air Force maintenance units (Charnes et al., 1985), hospitals (Grosskopf et al., 2001; Gruca and Nath 2001; Kirigia et al., 2001; Sola and Prior, 2001), university departments (Beasley, 1990), schools (Bessent and Bessent, 1980; Santos and Themido, 2001; Portela and Thanassoulis, 2001; Grosskopf and Moutray, 2001), counties (Raab and Lichty, 1997), as well as to compare industries and sectors (Sueyoshi and Goto, 2001; Navarro and Camacho, 2001; Murillo-Zamorano and Vega-Cervera, 2001; Mathijs and Swinnen, 2001), banks (Mukherjee et al., 2001; Sathye, 2001; Kuosmanen and Post, 2001; Hartman et al., 2001; Lin et al., 2009; Schaffnit et al., 1997), products and services (Doyle and Green, 1991; Hollingsworth and Parkin, 2001; Johnston and Gerard, 2001), computers (Doyle and Green, 1994), regulations (Piot-Lepetit et al., 2001; Gronli, 2001), strategic decision making (Demirbag et al., 2010), and technologies (Khouja, 1995; Shao and Lin, 2001; Ramanathan, 2001; Pare and Sicotte, 2001).

One of the fundamental assumptions of DEA is that all DMUs in the sample are *functionally similar* in the sense that all DMUs receive the same *number* and the same *type* of inputs and outputs. The compliance with this assumption is enforced by defining a common *DEA model*, according to which the evaluation of the relative efficiency of every DMU in the sample takes place. DEA treats a DMU as a collection of inputs and outputs, without any regard to the actual process by which conversion of inputs into outputs takes place; instead, the process of conversion is treated by DEA as a "black box" common to all DMUs in the sample. Another fundamental assumption of DEA is that a set of DMUs is *homogenous* in the sense that all DMUs are "alike" and thus directly comparable. Compliance with this important assumption of *homogeneity* of the sample is not enforced in DEA and usually resides under implicit purview of the decision maker. We suggest that two factors are important for the assumption of homogeneity of DMUs to hold. The first factor, *semantic homogeneity*, refers to the common meaning that is assigned to all DMUs in the sample by the decision maker. Compliance with this factor is straightforward. The second factor, *scale homogeneity*, refers to the *levels* of inputs and outputs of each DMU in the sample. In the absence of perfect scalability, the compliance with the second factor is problematic, for the decision maker must ensure that the levels of inputs and outputs are not affecting the functional similarity of DMUs in the sample.

However, in many situations the sample of DMUs is a sample of convenience. It is possible, under such circumstances, that an investigator conducts DEA using a sample consisting of DMUs that are functionally similar, semantically homogeneous, yet heterogeneous in terms of the levels of inputs and outputs, which would mean that the validity of the resulting relative efficiency scores would be questionable since the *homogeneity* assumption holds only partially.

We must note that we are not drawing attention of the reader to the obvious case where scale heterogeneity arises simply due to the differences in the scale of the transformation of inputs into outputs, with all other factors being constant. The scale heterogeneity could be easily countered through the use of a scaling factor. Instead, we are interested in the cases where scale heterogeneity arises because of a more complex, not clearly obvious, scaling pattern that makes creation of any accurate sub-categorization problematic. Let us consider an example of a comparison of a group of hospitals ranging in size from small to large. Clearly, the significant differences in the levels of inputs and outputs of such hospitals, whatever they might be, are important in themselves. However, it is also possible that such heterogeneity of the levels of inputs and outputs is reflective of the differences among the other important dimensions that are not accounted for by DEA model. In many cases, a large hospital is a very different enterprise from a standpoint of quality, complexity and technological sophistication than a small hospital.

Let us take this example further and consider that a smaller hospital A turned out to be relatively less efficient than a larger hospital B. Based on the results of the DEA of such *scale heterogeneous* sample of hospitals, the decision maker interested in improving efficiency of hospital A would face two possible options. The first option is to reduce the level of relative inefficiency of hospital A by improving the efficiency of the process by which inputs are converted into outputs. However, one cannot rule out that a lower level of relative efficiency of a hospital A is due to the comparatively lower level of inputs that it receives. Consequently, the second option is to reduce the level of relative inefficiency of hospital A by affecting the level of *scale heterogeneity*, i.e., to change the existing level of inputs. The question becomes, then, what option is to be pursued? Considering this, in our study we aim to investigate the following research question:

> *How to account for the differences in the relative efficiency scores of the DMUs in the sample in the presence of scale heterogeneity associated with a complex and non-obvious scaling pattern?*

Our inquiry relies on the assumption that the difference in the relative efficiencies of the DMUs in a *scale heterogeneous* sample could arise from the two sources. The first source reflects the difference in the *transformative capacity* of the DMUs, which reflects the "true" difference in the relative efficiencies of the DMUs in terms of the conversion of the set of inputs into the set of outputs. The second source reflects *scale heterogeneity* of the DMUs in the sample and is indicative of the difference in the *levels* of inputs and outputs associated with a complex and non-obvious scaling pattern.

We suggest that the assumption of homogeneity of a sample of DMUs should not be taken for granted, but explicitly tested for. The reader may note that this position is similar to what applies in statistical analysis where there is explicit testing of the normality assumption. Consequently, in the situation where an investigator needs to perform DEA, we suggest a multi-step methodology that is an extension of our previous work aimed at the increasing the discriminatory power of DEA (Samoilenko and Osei-Bryson, 2008).

## DESCRIPTION OF THE METHODOLOGY

The proposed methodology helps an investigator to address two potential problems that arise when conducting DEA. First problem is associated with a possible non-homogenous environment of DMUs (Dyson et al., 2001), which, due to its complexity, cannot be dealt with by simple inclusion of environmental variables in DEA model. For example, if we are to compare two hospitals in different states in the USA, in order to account for a possible heterogeneity of their environment we would need to include some sort of environmental variables reflecting the differences in social, political, environmental, legal, and cultural environments. Dyson et al. (2001) acknowledge that such environmental variables, especially in the service sector, could be difficult to identify, define, and measure. Furthermore, even if it is possible to completely account for the differences in the environment of DMUs and include the environmental variables into DEA model, this approach will result in a possibly significant increase in the number of inputs and output. This, in turn, would lead to the lower level of discrimination (Dyson et al., 2001). And while the authors suggested that this problem can be dealt with by increasing the number of DMUs in the sample, such increase could bring an additional source of non-homogeneity in the sample, which would have to be dealt with by inclusion of new types of environmental variables with all the consequences. The proposed use of NN in our methodology, on the other hand, allows an investigator to capture the impact of the specific environment for each logical sub-group without actually including any environmental variables in DEA model.

The second problem is associated with the available return-to-scale assumptions of DEA. Let us revisit our earlier example and compare two hospitals in terms of their relative efficiency of the production of revenues from the sale of a new type of service – a novel plastic surgery. The Product Life Cycle (PLC) model informs us that such *sales* curve is S-shaped (Sultan et al., 1990; Van den Bulte and Stremersch, 2004; Hauser et al., 2005). It is commonly accepted that the models that produce S-shaped curves (e.g., logistic curve model or Gompertz model) contain areas of increasing, constant, and decreasing return to scale. Using DEA to compare relative efficiency of two hospitals in such situation will require an investigator to impose two additional assumptions. First, it will be required to assume that our two hospitals are in the same phase of the *sales* curve (e.g., introduction, growth, maturity, or decline). Second, it will be required to assume that the estimate of the relative efficiency provided by DEA holds for the duration of the *sales* curve. Our approach utilizes NN to simulate the position of a DMU in more than one point on the *sales* curve, thus, the first assumption can be relaxed. Furthermore, by performing DEA of the simulated data we can obtain multiple scores of the relative efficiency for each DMU, which will allow us to relax the second assumption. Furthermore, using NN simulation in our methodology allows for avoiding DEA-related problems associated with the possible presence of economies and diseconomies of scale; consequently, we suggest adopting the basic CRS model.

The proposed methodology consists of five major steps that are summarized in Table 19.1; the description of each step follows. As previously mentioned above, the proposed methodology is an extension of the three-step methodology of Samoilenko

**TABLE 19.1**
**Proposed Methodology**

| **Step** | **Step 1** | **Step 2** | **Step 3** | **Step 4** | **Step 5** |
|---|---|---|---|---|---|
| **Purpose** | Evaluate the Scale Heterogeneity Status of the dataset | Determine the Relative Efficiency Status of each DMU | Generate a model of Transformative Capacity of each cluster | Obtain simulated sets of the outputs for each DMU in each cluster | Determine the sources of the Relative Efficiencies of the DMUs in the sample |
| **Data Set** | Complete sample | Complete sample | Clusters generated in Step 1 | Clusters generated in Step 1 | Complete sample |
| **Technique** | Cluster analysis (CA) | DEA | Neural networks (NN) | Neural networks (NN) | DEA |
| **Outcome** | One or more clusters | Scores of averaged relative efficiency for each cluster | "Black box" model of transformative capacity for each cluster | Simulated outputs for each cluster based on "Black box" models of other clusters | Scores of averaged relative efficiency for each cluster based on the original inputs and simulated outputs |

and Osei-Bryson (2008) that utilizes the CA and DEA as, correspondingly, the first and second steps. Consequently, in this section we provide only a brief description of the first two steps and focus our discussion on the remaining steps of the methodology proposed in this chapter.

## Description of Steps 3–5 of the Methodology

### Step 3: Generate a "Black Box" Model of Transformative Capacity of Each Cluster

*Description*

For each cluster $k$, NN induction is used to generate a "black box" model of the transformative capacity of DMUs in the given cluster. Let the model for cluster "$k$" be labeled $BBTM_k$. This involves using the set of input variables for each cluster as input nodes, and the set of output variables as output nodes of the NN. Then we train the NN in order to obtain the specific a *non-explanatory "Black box" model of the transformative capacity* by which the set of inputs is converted into the set of outputs; we call this *transformation model* of a given cluster.

### Step 4: Generate Simulated Sets of the Outputs for Each Cluster

For each cluster $k_1$, apply the black box transformative model $BBTM_{k1}$ to every other cluster $k_2$. The result is that for such that for each $DMU_{i(k2)}$ in a given cluster $k_2$, simulated outputs are generated based on the application of $BBTM_{k1}$ to the inputs of $DMU_{i(k2)}$.

### Step 5: Determine the Sources of the Relative Inefficiency of the DMUs in the Sample

For each pair of clusters ($k_1$, $k_2$):

a. Apply DEA to the original inputs of DMUs of cluster $k_2$ and the corresponding simulated outputs that resulted from $BBTM_{k1}$.
b. Calculate the average relative efficiency of cluster $k_2$ based on the its original inputs and simulated outputs.
c. Compare this simulated average relative efficiency of cluster $k_2$ to that of its actual average relative efficiency in order to determine if there is any difference in the transformative capacities of clusters $k_1$ and $k_2$.
d. Compare this simulated average relative efficiency of cluster $k_2$ to the actual average relative efficiency of cluster $k_1$ in order to determine if any difference in the average performances of clusters $k_1$ and $k_2$ is due, in part, to scale heterogeneity.

## Motivation for Steps 3 and 5 of the Methodology

### Motivation for Step 3

The third step of out methodology utilizes NN to create a model of the transformative capacity for each of the $k$ clusters identified in Step 1. We propose that modeling of

DEA scores can be used for the purposes of inquiring into the factors affecting relative efficiency, albeit indirectly. Despite the fact that the modeling of DEA scores is an often encountered approach (Hoff, 2006), it comes with the penalty of inevitable misspecification of the model according to which inputs are converted into outputs. It is reasonable to suggest, that the correct "white box" modeling of the DEA scores must have at least two pre-requisites. First, the investigator must know the model of transformation of inputs into outputs utilized by DEA. Second, it must be possible to re-specify correctly this known model within the data analytic technique that is going to be used for the purposes of modeling of DEA scores. The non-parametric nature of DEA is another point to consider; for use of parametric techniques, such as commonly used for this purpose Tobit regression, not only results in misspecification (Hoff, 2006), but also requires a compliance with the data normality assumption that often used in DEA sample of convenience may not satisfy.

As a result, a misspecification of the model is inevitable; for the "black box" approach of DEA to the process of transformation of inputs into outputs leaves investigator no chance of knowing, let alone specifying, the correct model. Keeping the mentioned above difficulties in mind, we decided instead to model a *new data set*, and then use DEA to obtain a new set of scores. The "black box" approach of NN to modeling complex unknown relationship in the data set fits well for this purpose, for we do not need to know and specify the relationship between the data ourselves.

We use the set of input variables for each cluster as input nodes, and the set of output variables as output nodes of the NN. Then we train the NN and obtain the specific to each cluster *transfer function* according to which the set of inputs is converted into the set of outputs; we call this *transfer function* a *non-explanatory "Black box" model of the transformative capacity* of a given cluster. Consequently, in the Step 3 we end up generating *k "Black box" models* corresponding to each of the *k* clusters identified in Step 1.

When the "Black box" models of the transformative capacity generated for every sub-set of the sample, we have successfully isolated the first factor influencing the relative efficiency of each DMU in the sample, namely, the efficiency of the transformation of the set of inputs into the set of outputs.

### Motivation for Step 5

The purposes of this step are: (1) to determine if the DMUs in a given cluster $k_2$ would have improved performance if they had utilized the transformative model of another cluster $k_1$; and (2) to determine if any difference is die to scale heterogeneity. For a given cluster this exploration is conducted using the transformative capacity model $\text{BBTM}_{k1}$ of every other cluster $k_1$ where $k_1 \neq k_2$.

Thus if Step 1 had resulted in two clusters, *Followers* and *Leaders*, during step 5 we would subject the original inputs and simulated outputs of the *Followers* that were generated in the four, as well as the original inputs and the simulated outputs of the *Leaders*, to DEA. Then we again calculate the average scores of the relative efficiency for every cluster and determine whether the averages of the *Followers* have improved. If this is the case, and the average relative efficiencies of the *Followers* have gone up, we have a reason to suggest that the disparity between the relative efficiencies of the *Leaders* and the *Followers* is due, in part, to the differences in their transformative capacities.

At this point, we need to determine a role of *scale heterogeneity* in the disparity of the levels of the relative efficiencies of the *Leaders* and the *Followers*. In order to do so we conduct DEA again, this time using the data set consisting of the original inputs and the outputs of the *Followers*, and the original inputs and simulated outputs of the *Leaders*. Once the scores of the relative efficiency have been obtained, we group them according to the cluster membership (i.e., *Followers* and *Leaders*), and average the relative efficiency scores for each group. If, after the comparison of the averaged relative efficiencies the *Leaders* still have a higher averaged relative efficiency score, then we have a reason to suggest that the disparity between the relative efficiencies of the *Leaders* and the *Followers* is due, in part, to scale heterogeneity. Meaning, even with the less efficient process of the transformative capacity, the *Leaders* are still capable of being more relatively efficient than the *Followers* are.

## ILLUSTRATIVE EXAMPLE

### DESCRIPTION OF THE ILLUSTRATIVE DATA SET

We test the proposed methodology in the context of the set of countries classified by International Monetary Fund as *Transition economies in Europe and the former Soviet Union*. Using archival data drawn from the *Database of World Development Indicators*, which is the World Bank's comprehensive database on development data, and the *Yearbook of Statistics*, published yearly by International Telecommunication Union (ITU), we aggregated the data on 18 economies for the period from 1993 to 2002. These 18 countries are Albania, Armenia, Azerbaijan, Belarus, Bulgaria, Czech Republic, Estonia, Hungary, Kazakhstan, Kyrgyz Republic, Latvia, Lithuania, Moldova, Poland, Romania, Slovak Republic, Slovenia, and Ukraine. As a result, we constructed the sample consisting of 180 data points, where each data point reflected a given TE per given year. While these economies do share a common classification, they also display some important differences in terms of their levels of economic development, state of infrastructure, business environment, etc. The research on the subject of the effect of investments in Information and Communication Technologies (ICT), such as Telecoms, on economic development in the context of TEs suggested that small number of countries (e.g., Poland, Hungary, Slovenia, and Czech Republic) were able to benefit from investments in ICT. The much larger number of TEs, however, falls short of demonstrating the positive impact of such investments on their economic development. Multiple research studies identified the level of investments in ICT (Murakami, 1997; Piatkowski, 2002) is one of the variables that impact the level of returns on investments.

Consequently, while keeping all the possible differences between 18 TEs in our sample in mind, it is reasonable to suggest that two factors could be responsible for the discrepancy of the effects of the investments in Telecoms on the level of returns on investment. The first factor is a level of investments in Telecoms and the second factor is a level of the efficiency of the transformation of investments in Telecoms into revenues.

To demonstrate our methodology in action, we formulate the following broad research problem:

> *How to determine the appropriate, empirically justifiable route by which TEs could improve their level of relative efficiency of the production of revenues from investments in Telecoms?*

In the context of the illustrative example and our methodology, the answer to this research problem involves answering the following questions:

1. Whether the 18 TEs display significant differences in terms of the levels of investments in Telecoms and revenues from Telecoms (Step 1).
2. Whether the sub-sets of the sample, which differ in terms of the levels of investments and revenues, also differ in terms of the relative efficiency of the production of revenues (Step 2).
3. Whether the sub-sets of the sample differ in terms of the processes of transformation on investments into revenues (Step 3 and Step 4).
4. Whether the relative inefficiency of poor performers is associated with the insufficient levels of investments or whether it is a result of inefficient processes of the transformation of investments into revenues (Step 5).

## Application of the Methodology on the Illustrative Data Set

For the DEA part of the methodology, we have identified a model consisting of six input and four output variables, presented in Table 19.2. A theoretical justification of the chosen model and the discussion regarding the choice of the variables can be found in Samoilenko (2008) and Samoilenko and Osei-Bryson (2008).

### Results of Step 1: Evaluate the Scale Heterogeneity Status of the Data Set

To perform CA, we employed a partitional approach to generate the maximum possible number of clusters (i.e., $k_{Max}$), followed by the application of an agglomerative clustering method to combine pairs of clusters until the specified minimum number of clusters (i.e., $k_{Min}$) is obtained. Given our interest in determining whether a set of

**TABLE 19.2**
**Variables Selected for the DEA Model**

| Input Variables of the DEA Model | Output Variables of the DEA Model |
|---|---|
| GDP per capita (in current US $) | Total telecom services revenue per telecom worker |
| Full-time telecommunication staff (% of total labor force) | Total telecom services revenue (% of GDP in current US $) |
| Annual telecom investment per telecom worker | Total telecom services revenue per worker |
| Annual telecom investment (% of GDP in current US $) | Total telecom services revenue per capita |
| Annual telecom investment per capita | |
| Annual telecom investment per worker | |

**TABLE 19.3**
**Membership of the 2-Cluster Solution**

| The *Followers* | The *Leaders* |
|---|---|
| **Albania (1993–2002)** | Bulgaria (2002) |
| **Armenia (1993–2002)** | **Czech Rep (1993–2002)** |
| **Azerbaijan (1993–2002)** | Estonia (1994–2002) |
| **Belarus (1993–2002)** | **Hungary (1993–2002)** |
| Bulgaria (1993–2001) | Latvia (1994, 1995, 1997–2002) |
| Estonia (1993) | Lithuania (1999–2002) |
| **Kazakhstan (1993–2002)** | **Poland (1993–2002)** |
| **Kyrgyzstan (1993–2002)** | **Slovenia (1993–2002)** |
| Latvia (1993, 1996) | Slovakia (1995–1998, 2000–2002) |
| Lithuania (1993–1998) | |
| **Moldova (1993–2002)** | |
| **Romania (1993–2002)** | |
| Slovakia (1993, 1994, 1999) | |
| **Ukraine (1993–2002)** | |

DMUs (i.e., 18 TEs) is a *scale heterogeneous*, we will use a user-specified threshold on outlier size to assess whether a given partition contains outlier clusters, and also use expert knowledge to further assess whether the partition is meaningful. A cluster will be considered an outlier if the percentage of the objects that it includes is less than $\tau_{\text{Outlier}}$ of the objects in the entire data set. We are not claiming that this is the only or always best approach, particularly since for a given data set it is never clear which approach is the most appropriate. Benefit of our approach, however, is that it allows for augmenting a context-independent solution with the context-dependent knowledge of a domain expert.

We used SAS Enterprise Miner (EM) to perform CA of the data set and we were able to come up with a solution that partitions our data set into two clusters (see Table 19.3). Based on the compiled information we can see, that while some of the TEs are "permanent residents" of one cluster, other TEs are "migrants", i.e., they change the cluster membership depending on a year.

### Results of Step 2: Determine the Relative Efficiency Status of Each DMU

To perform DEA, we have chosen an output-oriented model and used it under the conditions of constant return to scale (CRS), variable returns to scale (VRS) and non-increasing returns to scale (NIRS). Unlike in the case of the input-oriented model, the output-orientation does not concern itself with the efficient utilization of the inputs, but rather with the maximization of the outputs. Thus, it is probably reflective of the perspective of the investor, especially in the case when the primary goal is to obtain the maximum revenue.

We present the summarized results in Table 19.4. This approach of using average countries inefficiencies over the period of time, as well as averaged inefficiencies for a group of countries, is consistent with the approach of Arcelus and Arocena (2005).

**TABLE 19.4**
**DEA: Comparison of the Clusters based on the Output-oriented DEA Model**

| Relative Efficiency Score | *Leaders* | *Followers* | Difference | Difference % |
|---|---|---|---|---|
| CRS, Average | **1.94** | **2.54** | **−0.60** | **23.67** |

It turned out that the averaged relative efficiency of the second cluster is greater than of the first cluster; consequently, we labeled the first cluster as *Followers* and the second cluster as the "Leaders".

### Results of Steps 3 and 4: Generate Simulated Sets of the Outputs for Each Cluster Based on Black Box Models Transformative Capacity Processes

To conduct NN simulation of the outputs of our DEA model we used *Enterprise Miner* (a part of SAS 9.1 package by SAS Institute). For the purposes of this research, we used supervised mode of learning because the data set that we are going to use contains not only the inputs, but also the outputs. First, we used NN to simulate the outputs of the *Followers* based on the transformative capacity of the *Leaders*. We used the setting allowing us to use "Samples of data sets" for preliminary training. We also used "Average Error" as our model selection criterion; this setting chooses the model with the smallest average error for the validation data set. Other settings were: "none" as the number of preliminary runs and "standard back propagation" as a training technique. After the running of the model the convergence criterion was satisfied.

The process model diagram depicting stages involved into the simulation of the outputs is represented in Figure 19.1. Once the simulated outputs of the Followers were obtained, we used the same process to obtain the simulated outputs of the *Leaders* based on the transformative capacity of the *Followers*.

### Results of Step 5

Once the simulated outputs were obtained, they were substituted instead of the real outputs. After that, we have conducted the DEA again and obtained the new values

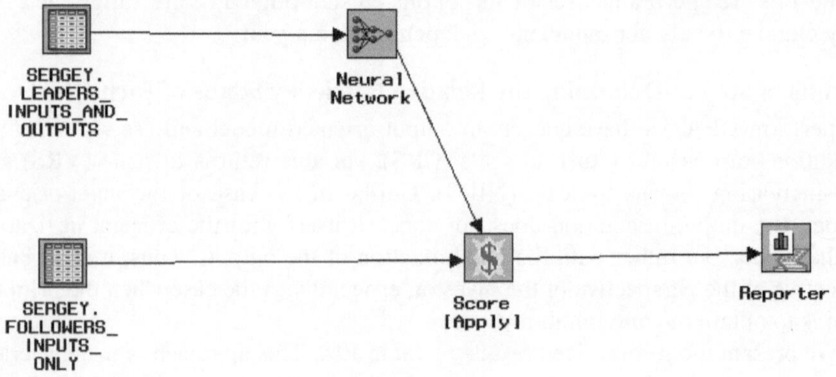


**FIGURE 19.1** The process model diagram of neural network simulation.

**TABLE 19.5**
**Summary of Step 5**

| Comparison of the DEA Scores based on | *Leaders*, CRS Averaged (Actual) | *Followers*, CRS Averaged (Simulated) | Difference | Difference % |
|---|---|---|---|---|
| The simulated data of the *Followers* and the actual of the *Leaders* | 2.04 | 1.62 | 0.42 | 25.62 |
| The simulated data of the *Leaders* and the actual of the *Followers* | 2.09 | 2.30 | -0.21 | 9.20 |

of relative efficiencies for the *Followers* and the *Leaders*. We adapted the approach used by Arcelus and Arocena (2000) and used averaged values of relative efficiency scores. The results are summarized in Table 19.5.

Based on the results of the DEA, we are able to establish that the *Followers* are capable of becoming more efficient than the *Leaders* in the case if they improve the level of the transformative capacity. Thus, at this point we can state that the lower level of the averaged relative efficiency of the *Followers* is, at least partially, a result of the inefficient processes of the revenue production. The results summarized in Table 19.5 also indicate, that the increase in the level of the inputs does not improve the level of relative efficiency of the *Followers*. Consequently, prior to increasing the level of the investments in Telecoms, the *Followers* must improve the effectiveness of the revenue production associated with the current level of the investments.

## DISCUSSION AND CONCLUSION

The non-parametric nature of DEA does not present a problem for the assessment of relative efficiency of technological artifacts, as long as they belong to the same domain.

When we attempt to compare level of performance of natural, social, or socio-technical entities we immediately end up on a fundamentally unstable ground, for there is very little we can do to ensure that subjects of our comparison are "alike". Let us consider some purely hypothetical examples. First, we compare a performance of two basketball players, *Star* and NotStar, in terms of their efficiency of converting minutes on the court into the points. *Star*, whose relative efficiency is higher, averages 35 minutes and 30 points per game, while *NotStar*, whose relative efficiency is lower, averages 5 minutes and 3 points per game. There are two possible explanations regarding the difference in the level of performance: first, *Star* is a better player than *NotStar*, second, *Star* is spending more time on the court. Now, how do we go about increasing relative efficiency of *NotStar*? Do we proceed by allowing him to play more minutes, or by trying to make him a better player first?

Similarly, how do investors should go about increasing the level of performance of the less successful baseball teams in Major League Baseball? Should investors provide incentives for a better performance, or should they start from investing more in the team's roster, hoping that the results will follow? Essentially, this is a type of a problem that our methodology allows us to address.

Let us consider an example of possible application of the proposed methodology, which is closer in spirit and relevance to the illustrative example that we used in this chapter. On November 20, 2001, the UN Secretary-General established, per request from the United Nations Economic and Social Council, an Information and Communication Technologies Task Force (ICTF). The purpose of this initiative was to provide a global dimension to the efforts in bridging of the "digital divide", to encourage digital opportunity, and to place ICT at the service of the development for all countries (Martinez-Frias, 2003). At this point, we know that in terms of the bridging digital divide the level of success from the investments in ICT varies greatly from country to country. However, we do not know whether the variation in the results is due to differences in the levels of the investments of ICT, or whether it is due to the capability of a given economy to transform the investments into the outcomes, i.e., bridging the digital divide. The answer to this question is very important, for it serves as one of the determinants of the investment strategy directed at the accomplishment of the goals of ICTF. We hope that we were able to demonstrate the possible contribution of the proposed methodology from practical standpoint.

We suggest that the proposed approach makes methodological contributions as well. First, our methodology allows for increasing the discriminatory power of DEA in the samples with the presence of heterogeneity. While traditional DEA alone categorizes DMUs in the sample as being relatively efficient or relatively inefficient, our approach allows for placing each DMU in one of the three categories. These categories are first, relatively efficient, second, relatively inefficient due to scale heterogeneity of the sample, and third, relatively inefficient due to transformative capacity. Second, our methodology allows for explicit acknowledging of the heterogeneity of the sample of DMUs, thus greatly expanding the domain of eligible for DEA DMUs. Finally, our approach allows for increasing prescriptive capabilities of DEA, providing a decision maker with distinct strategies regarding the increase of relative performance for each DMU in a non-homogenous sample.

We must acknowledge, however, that our research is not without its limitations. First, despite applying CA to evaluate a sample of DMUs for heterogeneity, our approach does not provide any strict criteria regarding to what constitutes heterogeneity of the sample. Consequently, because heterogeneity is a relative concept, and because the determination of heterogeneity often requires intimate knowledge of the problem domain, we declare this issue as being beyond the scope of our methodology and delegate it to reside under the purview of an investigator.

Second limitation of our study is associated with the assumption regarding the sources of relative inefficiency in the non-homogenous sample. We assume that heterogeneity of the sample *or* a transformative capacity of DMUs can cause relative inefficiency. However, it is possible that there is interplay between the two factors, where heterogeneity of the sample affects transformative capacity or, conversely, heterogeneity arises due to the differences in transformative capacity. Nevertheless, we hope that contributions of our study outweigh its limitations.

## ACKNOWLEDGMENT

Material in this chapter previously appeared in: Samoilenko, S., & Osei-Bryson, K. M. (2010). Determining sources of relative inefficiency in heterogeneous samples: Methodology using Cluster Analysis, DEA and Neural Networks. *European Journal of Operational Research*, *206*(2), 479–487.

## REFERENCES

Arcelus, F.J., Arocena, P., 2000. Convergence and productive efficiency in fourteen OECD countries: a non-parametric frontier approach. International Journal of Production Economics 66, 105–117.

Arcelus, F.J., Arocena, P., 2005. Productivity differences across OECD countries in the presence of environmental constraints. Journal of the Operational Research Society 56, 1352–1362.

Asmild, M., Paradi, J.C., Reese, D.N., Tam, F., 2007. Measuring overall efficiency and effectiveness using DEA. European Journal of Operational Research 178 (1), 305–321.

Beasley, J.E., 1990. Comparing university departments. Omega, International Journal of Management Science 18(2), 171–83.

Bessent, A., Bessent, W., 1980. Determining the comparative efficiency of schools through data envelopment analysis. Educational Administration Quarterly 16(2), 57–75.

Charnes, A., Clark, C.T, Cooper, W.W., Golany, B., 1985. A developmental study of data envelopment analysis in measuring the efficiency of maintenance units in the US Air Forces. Annals of Operations Research 2, 95–112.

Chen, C., van Dalen, J., 2010. Measuring dynamic efficiency: theories and an integrated methodology. European Journal of Operational Research 203(3), 749–760.

Chen, Y., Zhu, J., 2004. Measuring information technology's indirect impact on firm performance. Information Technology and Management, 5(1–2), 9–22.

Demirbag, M., Tatoglu, E., Glaister, K.W., Zaim, S., 2010. Measuring strategic decision making efficiency in different country contexts: a comparison of British and Turkish firms. Omega, 38, (1–2), 95–104.

Doyle, J, Green, R., 1994. Strategic choice and data envelopment analysis: comparing computers across many attributes. Journal of Information Technology 9(1), 61–69.

Doyle, J.R., Green, R.H., 1991. Comparing products using data envelopment analysis. Omega, International Journal of Management Science 19(6), 631–638.

Dyson, R.G., Allen, R., Camanho, A.S., Podinovski, V.V., Sarrico, C.S., Shale, E.A., 2001. Pitfalls and protocols in DEA. European Journal of Operational Research 132, 245–259.

Gillen, D., Lall, A., 1997. Developing measures of airport productivity and performance: an application of data envelopment analysis. Transportation Research (Part E) 33(4), 261–273.

Gronli, H., 2001. A comparison of Scandinavian regulatory models; issues and experience. The Electricity Journal 14(7), 57–64.

Grosskopf, S., Margaritis, D., Valdmanis, V., 2001. The effects of teaching on hospital productivity. Socio-Economic Planning Sciences 35(3), 189–204.

Grosskopf, S., Moutray, C., 2001. Evaluating performance in Chicago public high schools in the wake of decentralization. Economics of Education Review 20(1), 1–14.

Gruca, T., Nath, D., 2001. The technical efficiency of hospitals under a single payer system: the case of Ontario community hospitals. Health Care Management Science 4(2), 91–101.

Hartman, T., Storbeck, J., Byrnes, P., 2001. Allocative efficiency in branch banking. European Journal of Operational Research 134(2), 232–242.

Hauser, J., Tellis, J., Griffin, A., 2005. Research on innovation: a review and agenda for marketing science. Marketing Science, Forthcoming. Available at SSRN: http://ssrn.com/abstract=907230.

Hoff, A., 2006. Second stage DEA: comparison of approaches for modeling the DEA score. European Journal of Operational Research 181(3), 425–435.

Hollingsworth, B., Parkin, D., 2001. The efficiency of the delivery of neonatal care in the UK. Journal of Public Health Medicine 23(1), 47–50.

Johnston, K., Gerard, K., 2001. Assessing efficiency in the UK breast screening programme: does size of screening unit make a difference? Health Policy 56(1), 21–32.

Khalili, M., Camanho, A.S., Portela, M.C.A.S., Alirezaee, M.R., 2010. The measurement of relative efficiency using data envelopment analysis with assurance regions that link inputs and outputs. European Journal of Operational Research 203 (3), 761–770.

Khouja, M., 1995. The use of data envelopment analysis for technology selection. Computers and Industrial Engineering 28(1), 123–132.

Kirigia, J., Sambo, L., Scheel, H., 2001. Technical efficiency of public clinics in Kwazulu-Natal province of South Africa. East African Medical Journal 78(2), S1–13.

Kuosmanen, T., Post, T., 2001. Measuring economic efficiency with incomplete price information: with an application to European commercial banks. European Journal of Operational Research 134(1), 43–58.

Lin, T.T., Lee, C., Chiu, T., 2009. Application of DEA in analyzing a bank's operating performance. Expert Systems with Applications 36 (5), 8883–8891.

Martin, J., Roman, C., 2001. An application of DEA to measure the efficiency of Spanish airports prior to privatization. Journal of Air Transport Management 7(3), 149–57.

Martinez-Frias, J., 2003. The importance of ICTs for developing countries. Interdisciplinary Science Reviews 28(1), 10–14.

Mathijs, E., Swinnen, J., 2001. Production organization and efficiency during transition: an empirical analysis of East German agriculture. The Review of Economics and Statistics 83, 100–107.

Mukherjee, K., Ray, S., Miller, S., 2001. Productivity growth in large US commercial banks: the initial post-deregulation experience. Journal of Banking & Finance 25(5), 913–939.

Murakami, T. (1997). The Impact of ICT on Economic Growth and the Productivity Paradox. Available on-line at: http://www.tcf.or.jp/data/19971011_Takeshi_Murakami_2.pdf. Myers, D. (2004). Construction Economics. Spon Press, UK.

Murillo-Zamorano, L., Vega-Cervera, J., 2001. The use of parametric and non-parametric frontier methods to measure the productive efficiency in the industrial sector: a comparative study. International Journal of Production Economics 69(3), 265–275.

Navarro, J., Camacho, J., 2001. Productivity of the service sector: a regional perspective. Service Industries Journal 21(1), 123–148.

Pare, G, Sicotte, C., 2001. Information technology sophistication in health care: an instrument validation study among Canadian hospitals. International Journal of Medical Informatics 63(3), 205–23.

Pels, E., Nijkamp, P., Rietveld, P., 2001. Relative efficiency of European airports. Transport Policy 8(3), 183–92.

Piot-Lepetit, I., Brummer, B., Kleinhanss, W., 2001. Impacts of environmental regulations on the efficiency of arable farms in France and Germany. Agrarwirtschaft 50(3), 184–8.

Portela, M., Thanassoulis, E., 2001. Decomposing school and school-type efficiency. European Journal of Operational Research 132(2), 357–73.

Raab, R., Lichty, R., 1997. An efficiency analysis of Minnesota counties: a data envelopment analysis using 1993 IMPLAN input–output analysis. The Journal of Regional Analysis and Policy 27(1), 75–93.

Ramanathan, R., 2001. Comparative risk assessment of energy supply technologies: a data envelopment analysis approach. Energy 26(2), 197–203.

Samoilenko, S., 2008. Contributing factors to information technology investment utilization in transition economies: an empirical investigation. Information Technology for Development 14(1), 52–75.

Samoilenko, S., Osei-Bryson, K.-M., 2008. Increasing the discriminatory power of DEA in the presence of the sample heterogeneity with cluster analysis and decision trees. Expert Systems with Applications 34(2), 1568–1581.

Santos, J., Themido, I., 2001. An application of recent developments of data envelopment analysis to the evaluation of secondary schools in Portugal. International Journal of Services Technology and Management 2(1, 2), 142–160.

Sathye, M., 2001. X-efficiency in Australian banking: an empirical investigation. Journal of Banking and Finance 25(3), 613–630.

Schaffnit, C., Rosen, D., Paradi, J.C., 1997. Best practice analysis of bank branches: an application of DEA in a large Canadian bank. European Journal of Operational Research 98 (2), 269–289.

Shao, B., Lin, W., 2001. Measuring the value of information technology in technical efficiency with stochastic production frontiers. Information and Software Technology 43(7), 447–456.

Sola, M., Prior, D., 2001. Measuring productivity and quality changes using data envelopment analysis: an application to Catalan hospitals. Financial Accountability & Management 17(3), 219–245.

Sultan, F., Farley, J. U., Lehmann, D. R. 1990. A meta-analysis of applications of diffusion models. Journal of Marketing Research 27(1), 70–77.

Sueyoshi, T., Goto, M., 2001. Slack-adjusted DEA for time series analysis: performance measurement of Japanese electric power generation industry in 1984–1993. European Journal of Operational Research 133(2), 232–259.

Van den Bulte, C., Stremersch, S. 2004. Social contagion and income heterogeneity in new product diffusion: a meta-analysis test. Marketing Science 23(4), 530–544.

# 20 Exploring Context Specific Micro-Economic Impacts of ICT Capabilities

## INTRODUCTION

The role of information and communication technology (ICT) as a prominent enabler of sustainable growth has been noted (Greenhill, 2010). It is not surprising, therefore, that the mechanisms and links of the "ICT → impact of ICT" chain have been investigated with the purpose of improving the efficiency and effectiveness of the impact (Samoilenko, 2013, 2014). However, the specifics of implementation and operation of ICT-related resources differ greatly from industry to industry and from economy to economy. Resultantly, it is expected that the strength of the impact of ICT will vary among the economies of the world. This context dependency of this impact is worth investigating, for the results may provide valuable insights and practical implications to those economies that aim to improve their relative levels of effectiveness and efficiency of the impact of ICT. Theoretical frameworks that are consistent with grand theories and yet allow for discerning the differences in contexts could offer a foundation for studying an impact of ICT in a variety of diverse settings.

An important question to consider is: *Could an impact of ICT on development be made sustainable*? It is reasonable to assume that in the dynamic global business environment, the sustainability of the impact of ICT depends on the ability of economies to modify and manipulate a chain of intermediate links within the "investments in ICT → outcomes of investments in ICT" process. This assumption can be expressed in the form of the following general assertion:

> *Sustainability of the impact of a limited resource is influenced by the capability to manipulate the mechanism of the "resource → impact of the resource" process.*

Illustrative examples of this assertion are plentiful, but the most obvious one is continuous increase in the efficiency of automobile engines where the amount of gas in the tank could be considered a limited resource.

Another important question to consider is: *What are some of the conditions for ensuring a sustainable impact of ICT?* It is also reasonable to consider that due to the limited availability of the ICT-related inputs (e.g., investments in ICT), and subject to the law of diminishing returns, sustainability of the impact of ICT will be affected

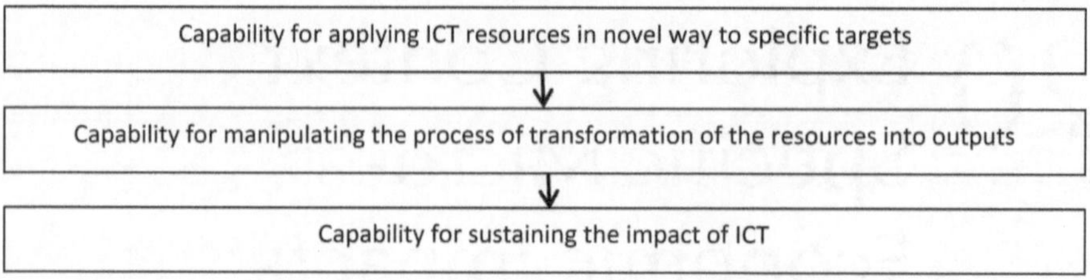


**FIGURE 20.1** Capabilities for the sustainable impact of information and communication technology.

by the ability to innovate. That is, the ability to apply available ICT resources in a novel way to specific targets that allows for "doing more with less". This consideration can be expressed as follows:

*Sustainability of the impact of a limited resource is influenced by the capability to apply the resource in novel ways.*

It should be noted that there are numerous examples of business functions where the limited resources of time, physical space, and money could limit the impact of ICT. These include: on-line learning, telecommuting, outsourcing, and offshoring.

Overall, the pre-requisites for a sustainable impact of ICT could be outlined as the chain of links depicted in Figure 20.1.

Currently there is no framework allowing for assessing the impact of ICT-related capabilities on specific economic targets that contribute to the overall macroeconomic impacts (e.g., GDP). The presence of such a framework would be of benefit to investigators and policy makers alike, for it would allow for assessing the impact of investments in ICT on ICT capabilities, and the impact of ICT capabilities on economic outcomes. The general model of "Investments in ICT – Impact of Investments in ICT" adopted in this study is depicted in Figure 20.2.

The justification for this general model is intuitive: investments in ICT could not produce anything on their own (e.g., it is simply an amount of money in the bank, or a line on a balance sheet), instead, investments must be converted into capabilities that allow for producing a desired impact (e.g., money allocated for grocery must be converted into purchased food and then into a capability of an individual to work). Furthermore, no capability directly impacts any measure of macroeconomic outcomes (e.g., a capability of an individual to work is not directly converted into a paycheck), instead, a capability impacts a more or less specific target (e.g., for it

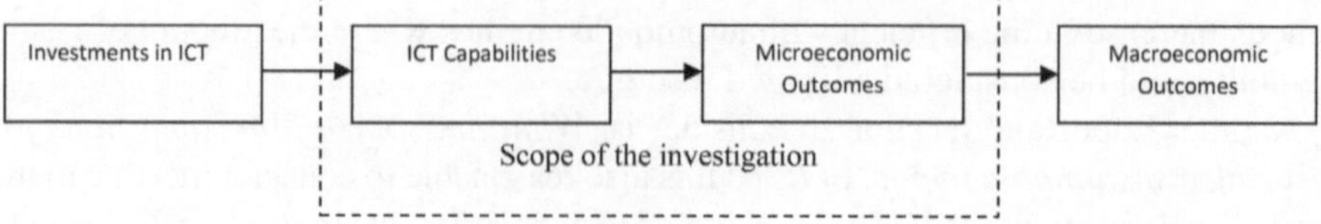


**FIGURE 20.2** General model of "Investments in ICT – Impact of Investments in ICT".

is a capability for doing something) that is later aggregated into macroeconomic outcomes. The chain of links presented in Figure 20.2 illustrates the scope and the limitation of the current study: we limit our inquiry to the economic aspect of the impact of ICT, and we operate within the boundaries of the impact of ICT capabilities on microeconomic outcomes. For all intents and purposes we use the construct "ICT Capabilities" to represent a *state of ICT* in a given context.

This chapter aims to contribute to the existing body of knowledge in the area of ICT for Development (ICT4D) by accomplishing its overall goal, stated as follows:

> *The main objective of this investigation is to develop and test a framework allowing for investigating the context-specific impacts of ICT capabilities on microeconomic outcomes.*

Because our study considers the impacts of various aspects of ICT capabilities on specific microeconomic targets, we argue that the obtained insights and their implications would contribute to an understanding of how to increase the level of sustainability of the impact of ICT, as well as to pinpointing where innovations in the area of application of ICT would be of the most benefit. Specifically, in this investigation we present a methodology for studying the impact of ICT capabilities on microeconomic outcomes, and apply this methodology to 5 groups of 24 economies of the world, where the economies are assigned to the groups in accordance with a classification scheme of the International Monetary Fund as of September 2011.

The five groups are: the Advanced Economies, Central and Eastern Europe; the Commonwealth of Independent States and Mongolia, the Middle East and North Africa; and Sub-Saharan Africa (SSA). The choice of the groups was driven by the goal of creating diversity of contextual representation for the empirical component of our investigation. The issue of the context of the capabilities and impacts of ICT is an important one, for there is a consensus that the setting greatly impacts the efficiency and the effectiveness of the translation of investments into ICT capabilities, and ICT capabilities into socio-economic impacts. Consequently, it is reasonable to expect that context-specific ICT-related policies would be more effective than generic ones. By comprising our sample of a heterogeneous set of five groups, we created a testbed that allows for discerning the specificity of the contexts and deriving insights that could be used in crafting of local policies.

To proceed with the investigation, we need to operationalize the *ICT Capabilities* and *Microeconomic Outcomes* constructs. In this study we represent *ICT Capabilities* via the *Networked Readiness Index* (NRI) – a metric for monitoring the development of countries with regards to the state of technology, where an increase in the value of NRI is indicative of the increase of the impact of ICT on innovation and productivity (Dutta & Mia, 2011). The NRI is comprised of three subindexes – *Environment*, *Readiness*, and *Usage*. We represent *Microeconomic Outcomes* via three precursors of GDP – *State of the Labor Market*, *International Trade*, and *Economic Well-Being* of the population (Samoilenko, 2013, 2014). This results in a framework with six constructs, where each of the three input-side constructs is associated with each of the three output-side constructs.

Resultantly, in this study we look at the nine paths representing impacts of three constructs of NRI – *Environment*, *Readiness*, and *Usage*, on three precursors of GDP – *State of the Labor Market*, *International Trade*, and *Economic Well-Being* of the population. Given five economy groups and nine "ICT Capabilities → Microeconomic Outcomes" paths, we aim to answer the following research questions:

1. *RQ1: What are the specific characteristics of each group with regards to ICT and the impacts of ICT?* This question allows us to test whether the proposed framework allows for discerning context-specificity of the setting.
2. *RQ2: What areas of ICT may require innovative applications of the available resources for each group?* This question tests whether the framework allows for identifying those ICT capabilities that are lagging within a given context.
3. *RQ3: What are the areas of strength and weakness in terms of the efficiency of the "ICT Capabilities → Microeconomic Outcomes" paths for each group?* The last question tests the capability of the framework to identify those mechanisms of transformation of ICT capabilities into macroeconomic outcomes that need to be adjusted.

Answering these questions within the context of the developed framework allows us to address the general research question of this study, which is formulated as follows:

> *What are the context-specific factors that differentiate various groups of economies in terms of the impact of ICT?*

This general research question aims to test whether the proposed framework achieved the overall goal of this study. The significance of the answering this question is intuitive, for

> *Knowing the context – and the impact-specific ICT factors allows for formulating precise policies and implementing custom-tailored practical solutions in the area of ICT4D.*

The justification for our inquiry is also straightforward, because

> *The diversity of the context of application of ICT4D precludes policy – and decision makers from formulating effective generic solutions.*

## THEORETICAL FRAMEWORK AND THE RESEARCH MODEL

A model of neoclassical growth accounting (Solow, 1957), widely used by researchers to estimate the contribution of ICT to the macroeconomic bottom line of developed, developing and transition economies (Oliner & Sichel, 2000; Schreyer, 2000; Daveri, 2000; Jorgenson & Stiroh, 2000; Whelan, 2000; Hernando & Nunez, 2002, Samoilenko & Osei-Bryson, 2008, Samoilenko & Ngwenyama, 2011, Samoilenko, 2013, Samoilenko & Osei-Bryson, 2016), serves as a grand theory that supports our

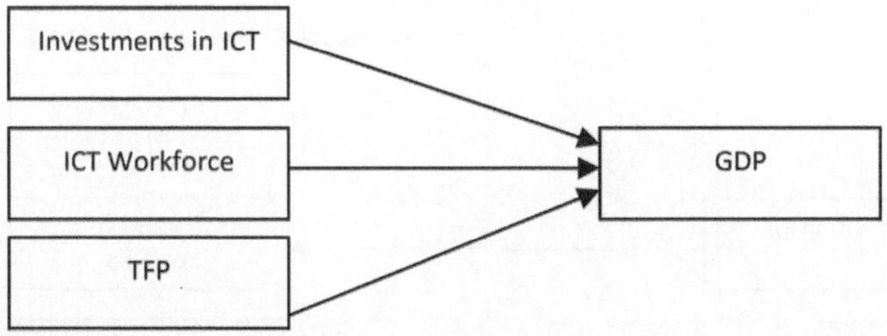


**FIGURE 20.3** Macroeconomic impact of investments in ICT based of the neoclassical growth accounting model.

investigation. The framework allows for decomposing the overall growth of an economy into the contributions from various inputs. A common formulation of neoclassical production function is:

$$Y = f(A, K, L), \tag{20.1}$$

where $Y$ = measure of economic output (most often in the form of GDP); $K$ = measure of capital, an endogenous variable explaining part of $Y$; $L$ = measure of labor, also an endogenous variable explaining part of $Y$; and $A$ = total factor productivity (TFP), an exogenous, unexplained by the endogenous components of $Y$.

While this model offers investigators a theoretically sound platform to conduct research, it clearly lacks details regarding mechanisms by which investments in ICT are transformed into GDP. It would be advantageous to identify some of the constructs, or precursors, that are impacted by the investments in ICT and which are, in turn, impact GDP. In the context of ICT4D, the neoclassical growth accounting model could be depicted as presented in Figure 20.3.

Based on the model of neoclassical growth accounting, a framework linking ICT to state of the labor market, international trade, and financial well-being of the population (see Figure 20.4) was developed by Samoilenko (2013).

While this framework offers a more detailed perspective on the paths by which investments in ICT contribute to GDP, it does not deal with the context of these investments. But the context, the environment where investments are made, is very important, because the same amount of investments could produce different degree

**FIGURE 20.4** Framework of Samoilenko (2013) consistent with the neoclassical growth accounting.

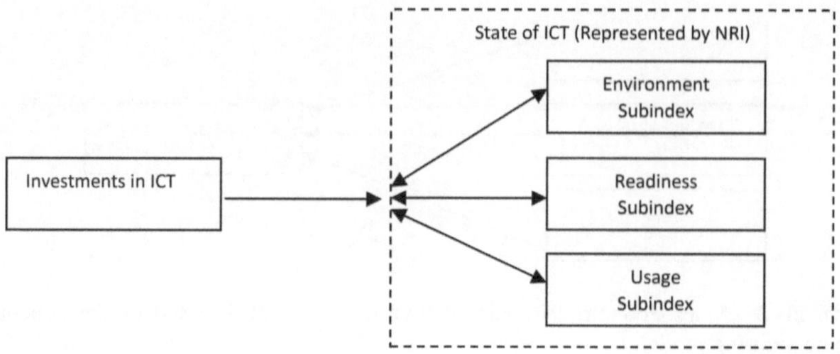


**FIGURE 20.5** Relationship between investments in ICT and the state of ICT represented by NRI.

of impacts in different locales – 5-million-dollar E-Government initiative would bring different results in Singapore than it would in Ghana.

A theoretical framework of representation of the state of ICT in the form of *Network Readiness Index* (NRI), consisting of environment, usage, and readiness subindexes, was first outlined in 2003 within the Global Information Technology report (Dutta & Jain, 2003). Since the introduction of NRI there were changes in regard to included variables and the methodology for computing the rankings (Dutta, Bilbao-Osorio, & Gieger, 2012), but three sub-indexes – the drivers of ICT impact – has remained components of NRI since its introduction. There is a clear link between investments in ICT and the state of ICT; consequently, we could represent the relationship between investments in ICT and the subindexes of NRI model as depicted in Figure 20.5.

Overall, the framework of this inquiry, as presented in Figure 20.6, is an integration of the NRI framework and the framework of the microeconomic impact of investments in telecoms (Samoilenko, 2013).

The important assertion of this study is stated as follows:

> *Investments in ICT contribute to the current state of ICT, and the state of ICT reflects, and is reflected by, ICT Capabilities.*

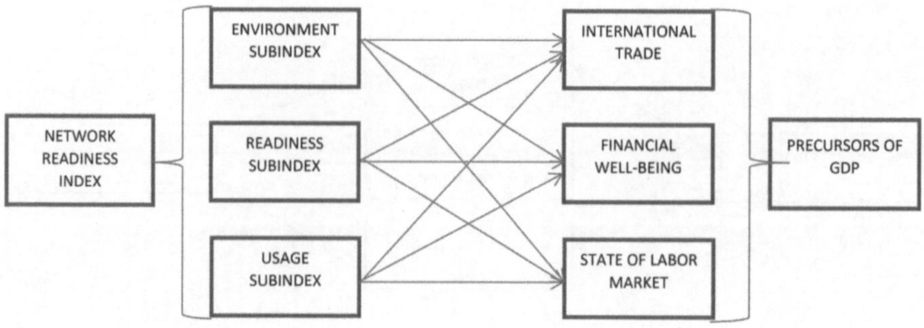


**FIGURE 20.6** Integrated theoretical framework of the study.

The framework is general enough for a great variety of contexts – values of NRI are available for over 145 economies, and the proposed precursors of GDP are common to the economies of the world. In the same time, the framework is adaptable – a decision maker may choose the representation of the precursors of GDP that better fits the local context.

The benefits of using the proposed framework in this study are two-fold. First, it allows for investigating the efficiency of mechanisms of transformation of the ICT-related resources into the microeconomic outcomes-precursors of the GDP. Specifically, the framework allows for investigating efficiency of target-specific impacts of ICT (e.g., *Does ICT readiness have a greater efficiency of the impact on the Labor Market than on International Trade*?). Additionally, the framework allows for identifying areas of application of ICT that may require innovative application of ICT resources (e.g., *What could be done in the area of ICT usage to improve the efficiency of its impact on Financial Well-Being of the population?*).

## THE METHODOLOGY OF THE STUDY

Our methodology consists of two phases: Data Envelopment Analysis (DEA) and Decision Tree (DT) analysis.

### PHASE 1: APPLICATION OF DATA ENVELOPMENT ANALYSIS (DEA)

The purpose of DEA is to identify relatively efficient economies with regard to the process of conversion of DEA inputs into DEA outputs. This analysis involves nine DEA models (*Input#.Output#*), where each DEA model refers to one of the "input–output" paths of the proposed framework (see Table 20.1).

**TABLE 20.1**
**Description of Input and Output Components of Nine DEA Models**

| Label | DEA Inputs (ICT Capabilities) | Label | DEA Outputs (Microeconomic Outcomes) |
|---|---|---|---|
| I1 | 1. Market Environment<br>2. Political and Regulatory Environment<br>3. Infrastructure Environment | O1 | 1. Imports of goods and services (% of GDP)<br>2. Exports of goods and services (% of GDP) |
| I2 | 1. Individual Readiness<br>2. Business Readiness<br>3. Government Readiness | O2 | 1. Health expenditure, private (% of GDP)<br>2. International tourism, expenditures (% of GDP) |
| I3 | 1. Individual Usage<br>2. Business Usage<br>3. Government Usage | O3 | 1. Labor force participation rate, male (% of male population ages 15+) (modeled ILO estimate)<br>2. Labor force participation rate, female (% of female population ages 15+) (modeled ILO estimate) |

The representation of the DEA inputs is based on the representation of the sub-indexes of NRI used by Global Information Technology reports 2011 and 2012. The representation of the DEA outputs (Microeconomic Outcomes) is based on the constructs developed in Samoilenko (2013, 2014). Each DEA model is labeled according to "Input#.Output#" naming convention (e.g., ***I1.O1*** through ***I3.O3***). According to this labeling scheme DEA model ***I3.O1***, for example, represents a model of "[Individual usage, Business usage, Government usage] → [Imports of goods and services, Exports of goods and services]".

We now describe the steps of this phase, beginning with the question that it aims to address followed by an outline of what is done in the given step.

## Phase 1, Step 1

1. *What ICT inputs (ICT Capabilities) would benefit the most from innovative application of ICT?*

For each *Input/Output* path of the framework, construct and execute a corresponding input-oriented DEA model for each *Economy Group* to generate *Relative Efficiency* and Malmquist Index (MI) scores. For each *Economy Group*, use the generated *Relative Efficiency* scores to identify those countries that are *relatively inefficient* under the assumption of input-orientation and thus would benefit the most from an innovation.

## Phase 1, Step 2

2. *Which ICT Capability/Microeconomic Outcome path would benefit the most from the changes to its input–output transformation process?*

For each *Input/Output* path of the framework, this question can be answered by analyzing the MI scores in order to identify the DMUs (e.g., countries, Economy Groups) that exhibited negative growth in productivity (i.e., MI < 1) and so would benefit the most from changes to their input–output transformation process. Therefore, for each *Input/Output* path and Economy Group, generate Malmquist Index (MI) scores, and evaluate whether each corresponding MI is less than 1.

## Phase 1, Step 3

3. *What are the areas of strength and weakness of DMU in terms of the efficiency of the "ICT capabilities → Microeconomic Outcomes" paths for each group?*

This question can be addressed using two criteria that are based on the MI scores and the values of its components, the *Efficiency Component* (EC) and the *Technology Component* (TC). For a given path, if the MI > 1 for the given economy then it has exhibited growth in productivity. If the corresponding EC > TC then this growth in productivity is due to improvement in *Efficiency*; while if the corresponding TC > EC then this growth in productivity is due to improvement in the application of the *Technology*.

## Phase 2: Decision Tree-Based Analysis

The second step of the methodology involves DT analysis. The purpose of the DT induction is to identify the input variable that is selected for the top-level partitioning of the data set. Specifically, DT analysis will allow for answering the following questions described below in Steps 1 and 2 of this phase:

### Phase 2, Step 1

1. *What are the differences between the groups of economies in our sample in terms of ICT-related factors and outcomes for each year?*

To answer this question, we use DT induction to generate separate DTs of depth 2 (i.e., each rule will have at most two predictor variables) for the years 2010 and 2011. It should be noted that our input data set contains an *Economy Group* variable that indicates the *Economy Group* associated with the given row of the data set. In this application of DT induction, the *Economy Group* variable is used as the target variable, and the component variables associated with *ICT Capabilities* (see Table 20.2) are used as the potential predictor variables. The reader may recall the DT induction process involves recursive partitioning of the data set based, where the split of the potential predictor variable that provides the top value of the relevant splitting measure is used for the first partitioning of the data set into its first-level sub-sets, and that subsequent partitioning of these sub-set involves the same approach. The resulting DT will thus have branches that are associated with the splits of the given variable, and nodes that provide the associated relative frequencies for the *Economy Groups*. Our aim in using DT induction is to identify the nodes of the DT in which one of the *Economy Groups* has a relative frequency of at least 50%. This will then allow us to identify the range of values of the top *ICT Capabilities* component variables that are associated with each *Economy Group*. It should, however, be noted that based on the actual data it is possible that for a given Economy Group there may not be any node of the resulting DT for which it has a relative frequency of at least 50%. This would indicate that sufficiently strong differentiating factors were not identified.

**TABLE 20.2**
**Groups of Countries (Based on IMF Classification of 2011) and Membership of Each Group**

| Advanced Economies | Central and Eastern Europe | Commonwealth of Independent States and Mongolia | Middle East and North Africa | Sub-Saharan Africa |
|---|---|---|---|---|
| Estonia | Hungary | Armenia | Morocco | Kenya |
| Slovenia | Latvia | Kazakhstan | Tunisia | Ghana |
| Czech Republic | Poland | Moldova | Algeria | Senegal |
| Slovak Republic | Lithuania | Kyrgyz Republic | Oman | Namibia |
| Spain | Montenegro | Tajikistan | | Nigeria |

### Phase 2, Step 2

2. *What are some of the differentiating characteristics of the Relatively Efficient and Inefficient economies?*

To answer this question, we use DT induction to generate separate DTs of depth 2 (i.e., each rule will have at most two predictor variables) for each of our DEA models. It should be noted that our input data set is extended to contain an *Efficiency_Class* variable that indicates whether the Economy Group associated with the given row of the data set is relatively *Efficient* or relative *Inefficient.* In this application of DT induction in this phase, the *Efficiency_Class* variable is used as the target variable, and the component variables associated with *ICT Capabilities* and *Microeconomic Outcomes* (see Table 20.2) are used as the potential predictor variables. Here our aim in using DT induction is to identify the nodes of the DT in which one of the *Efficiency_Class*es has a relative frequency of at least 50%. This will then allow us to identify for each of our DEA Models, the range of values of the top *ICT Capabilities* and *Microeconomic Outcomes* component variables that are associated with each *Efficiency_Class.* It should, however, be noted that based on the actual data it is possible that for a given *Efficiency_Class* there may be not be any node of the resulting DT for which it has a relative frequency of at least 50% which would indicate that sufficiently strong differentiating factors based were not identified.

## DESCRIPTION OF THE DATA

We obtained the data from two sources – the database of the World Development Indicators (WDI) and Global Information Technology Reports of 2010 and 2011. A two-year period, while may not being sufficient for the purposes of performing an in-depth time series data analysis, is sufficient for the purposes of illustrating and evaluating the proposed framework and associated methodology. Overall, we compiled the data on 24 economies of the world representing five groups according to the classification of the International Monetary Fund 2011 (Nielsen, 2011). Membership of each group is provided in Table 20.2. The NRI framework does not split the set of the countries into the various sub-groups when assigning scores to the subindexes. Consequently, our partitioning of the sample of 24 countries into 5 groups is logical in nature.

The limited number of the economies of this study is due to exploratory nature of the investigation – once the framework is developed and tested we will increase the sample size and conduct the follow-up study in a large context.

## RESULTS OF THE DATA ANALYSIS

### Results from Phase 1: Application of Data Envelopment Analysis (DEA)

### Phase 1, Step 1

We start by addressing the second research question RQ2:

- *What areas of ICT may require innovative applications of the available resources for each group?*

**TABLE 20.3**
**Average Relative Efficiency Score for Each Model**

| | Model | | | | | | | | |
|---|---|---|---|---|---|---|---|---|---|
| **Year** | **I1.O1** | **I1.O2** | **I1.O3** | **I2.O1** | **I2.O2** | **I2.O3** | **I3.O1** | **I3.O2** | **I3.O3** |
| 2010 | ***0.60*** | *0.66* | *0.62* | *0.61* | *0.67* | *0.63* | *0.80* | *0.83* | *0.84* |
| 2011 | ***0.55*** | *0.66* | *0.58* | *0.60* | *0.63* | *0.63* | *0.78* | *0.87* | *0.82* |

In order to do so, for each model we first calculate Relative Efficiency score for each economy for each year, which then allows us to identify the model with the lowest average Relative Efficiency score (see Table 20.3). In our case, for both years, it is a model *I1.O1*, and the corresponding ICT Capability is *Environment* (label *I1*).

The results allow us to answer RQ2 as follows:

*Market Environment, Infrastructure Environment, and Political and Regulatory Environment are the areas of ICT Capabilities that may require innovative application of the available resources for each group.*

## Phase 1, Step 2

We next addressed the question:

*Which ICT Capability/Microeconomic Outcome path would benefit the most from the changes to the input–output transformation process?*

In order to do so, we first generated the Malmquist Index (MI) scores for each economy for each model (for details see Table 4A of the Appendix), then calculated the average MI score for each model (see Table 20.4), following which we assessed which of these scores was less than or equal to 1. Given our results displayed in Table 20.4, the answer to the second question for the period of 2010–2011 is:

*The ICT Capabilities to Financial Well-Being paths would benefit the most from the changes in the Input–Output Transformation Process.*

An interesting pair of related issues are:

- What is the economy group that should be used as a benchmark of efficiency for a given model?
- What is the economy group that shows the greatest improvement for a given model?

**TABLE 20.4**
**Average Malmquist Index (MI) Score for Each Model**

| | Model | | | | | | | | |
|---|---|---|---|---|---|---|---|---|---|
| **Period** | **I1.O1** | **I1.O2** | **I1.O3** | **I2.O1** | **I2.O2** | **I2.O3** | **I3.O1** | **I3.O2** | **I3.O3** |
| 2010–2011 | *1.11* | ***0.97*** | *1.03* | *1.10* | ***0.95*** | *1.02* | *1.10* | ***1.00*** | *1.01* |

**TABLE 20.5**
**Benchmark Group for Each Model**

| Model | Benchmark Economy Group 2010 | 2011 |
|---|---|---|
| **I1.O1:** Environment → Trade | Advanced Economies (AE) | AE |
| **I1.O2:** Environment → Financial Well-Being | AE | AE |
| **I1.O3:** Environment → State of Labor Market | AE | AE |
| **I2.O1:** Usage → Trade | Comm. of Ind. States (CIS) | CIS |
| **I2.O2:** Usage → Financial Well-Being | CIS | CIS |
| **I2.O3:** Usage → State of Labor Market | CIS | CIS |
| **I3.O1:** Readiness → Trade | AE | AE |
| **I3.O2:** Readiness → Financial Well-Being | AE | AE |
| **I3.O3:** Readiness → Financial Well-Being | AE | AE |

In order to make this determination of the benchmark group for efficiency we calculate the average relative efficiency score of each economy group, per model per year. For each model, the economy group that provides the highest average relative efficiency score is then selected as the benchmark group for that year. Table 20.5 identifies the benchmark economy group for each model for each year.

## Phase 1, Step 3

With regards to our third research question RQ3 (i.e., *What are the areas of strength and weakness in terms of the efficiency of the "ICT Capabilities → Microeconomic Outcomes" paths for each group*), evaluation of the strengths and weaknesses can be performed using two criteria: (1) whether a given economy exhibited growth in productivity for a given path? (e.g., is MI > 1 or not?); and (2) what is the dominant source of growth in productivity? (e.g., which component is greater in value, EC or TC?). The information presented in Tables 20.6 and 20.7 could be used to answer these questions.

**TABLE 20.6**
**Growth (+)/Decline (–) in Productivity per Economy Group per Model**

| Economy Group | I1.O1 | I1.O2 | I1.O3 | I2.O1 | I2.O2 | I2.O3 | I3.O1 | I3.O2 | I3.O3 |
|---|---|---|---|---|---|---|---|---|---|
| AE | + | – | + | – | – | – | + | – | – |
| CEE | + | – | + | + | – | – | + | + | + |
| CIS | + | – | + | + | + | + | + | – | + |
| MENA | + | – | + | + | + | + | + | + | + |
| SSA | + | – | – | + | – | – | + | – | + |

**TABLE 20.7**
**Economy Group with Greatest Change in Productivity per Model**

| Model | Economy Group | Dominant Source |
|---|---|---|
| **I1.O1:** Environment → Trade | Advanced Economies | TC |
| **I1.O2:** Environment → Financial Well-Being | Middle East and North Africa (MENA) | TC |
| **I1.O3:** Environment → State of Labor Market | Central and Eastern Europe (CEE) | TC |
| **I2.O1:** Usage → Trade | Comm. of Ind. States (CIS) | EC |
| **I2.O2:** Usage → Financial Well-Being | CIS | TC |
| **I2.O3:** Usage → State of Labor Market | CIS | EC |
| **I3.O1:** Readiness → Trade | CEE | TC |
| **I3.O2:** Readiness → Financial Well-Being | CEE | EC |
| **I3.O3:** Readiness → Financial Well-Being | MENA | EC |

Information summarized in Table 20.6 allows for answering RQ2 as follows:

*With regard to the efficiency of the "ICT capabilities → Microeconomic Outcomes" paths, for each group, the areas of strength are associated with the growth in productivity and the areas of weakness with the decline in productivity.*

This allows us to conclude *that the proposed framework successfully passed the test of the capability to identify those mechanisms of transformation of ICT capabilities into macroeconomic outcomes that need to be adjusted.*

## Results from Phase 2 – Decision Tree (DT) Based Analysis

We now focus on the first research question RQ1, namely:

*What are the specific characteristics of each group with regards to ICT and the impacts of ICT?*

To address this question we executed *Phase 2. Step 1* (see "Phase 2: Decision Tree Based Analysis" section); the corresponding results are displayed in Table 20.8. The reader may note that differentiating factors were identified for all our economy groups with the with the exception of the Middle East and North Africa (MENA) group.

Given the information displayed in Table 20.9 the answer to RQ1 is as follows:

*The difference between the groups of economies in our sample could be expressed in terms of the differences with regards to the following ICT Capabilities*

- *Market Environment and Individual Usage in 2010, and*
- *Market Environment and Infrastructure Environment in 2011.*

**TABLE 20.8**
**Differences between the Groups of Economies in Terms of ICT Capabilities**

| Year | Group | Differentiating Factors | Classification |
|---|---|---|---|
| 2010 | Advanced Economies (AE) | MarketEnv ≥ 4.355 & IndUse ≥ 3.5 | 100% of AE |
| | Central and Eastern Europe (CEE) | MarketEnv < 4.355 & IndUse ≥ 3.5 | 100% CEE |
| | Comm. of Ind. States (CIS) | MarketEnv < 3.925 & IndUse < 3.5 | 83% of CIS |
| | Middle East and North Africa (MENA) | N/A | N/A |
| | Sub-Saharan Africa (SSA) | MarketEnv ≥ 3.925 & IndUse < 3.5 | 63% of SSA |
| 2011 | AE | InfraEnv > 3.65 | 100% of AE |
| | CEE | InfraEnv > 3.65 | 100% of CEE |
| | CIS | MarketEnv < 3.85& InfraEnv < 3.65 | 83% of CIS |
| | MENA | N/A | N/A |
| | SSA | MarketEnv ≥ 3.85& InfraEnv < 3.65 | 63% of SSA |

It can be seen that the information that results from the execution of *Phase 2. Step 1* could be used by policymakers/decision-makers for formulating improvement strategies. For example, if an economy-member of the SSA group aims to benchmark a performance of an economy-member of the Advanced Economies (AE) group, then the results show that in regard to *ICT Capabilities* it could be wise to direct attention to *Individual Usage*, for this is the area where two groups of economies are clearly different. This may mean increasing the level of ICT investments in this area, or applying existing levels of investments in innovative way.

We now focus on the second research question of this phase:

*What are some of the differentiating characteristics of the relatively efficient and relatively inefficient economies?*

**TABLE 20.9**
**DT Analysis: Characteristics of Efficient and Inefficient Economies, per Model**

| Model | Characteristics of Efficient Economies | Characteristics of Inefficient Economies |
|---|---|---|
| Environment → Trade | N/A | N/A |
| Environment → Income | BusRead ≥ 4.245 | BusRead < 4.245 |
| Environment → Labor | Exports ≥ 43.4345 | Exports < 43.4345 |
| Readiness → Trade | IndRead ≥ 4.51 | IndRead < 4.51 |
| Readiness → Income | GovRead ≥ 4.345 | GovRead < 4.345 |
| Readiness → Labor | IndRead ≥ 4.545 & BusRead <4.005 | IndRead < 4.545 & BusRead < 4.005 |
| Usage → Trade | N/A | N/A |
| Usage → Income | Tourism ≥ 0.01515 & BusRead ≥ 4.04 | Tourism≥0.01515 & BusRead <4.04 |
| Usage → Labor | BusRead < 4.245 | BusRead ≥ 4.245 |

To address this question we executed *Phase 2. Step 2* (see "Phase 2: Decision Tree Based Analysis" section); the corresponding results are displayed in Table 20.9. The reader may note that differentiating factors were identified for all our DEA models with the with the exception of the "Environment → Trade" and "Usage → Trade" models.

Let us consider usefulness of the information from the execution of *Phase 2. Step 2.* For example, in the case of the model *Environment → Income* (which represents "[Market environment, Political and regulatory environment, Infrastructure environment] → [Private health expenditure, Expenditures on International tourism]"), one would expect that a relatively inefficient economy in regard to this path should concentrate its resources on *Environment*-related capabilities, or *Income*-related microeconomic outcomes, or, on the process of transformation between the two. Instead, the results show that the emphasis should be placed on *Business Readiness* capability – this may indicate a presence of complementarity between the *Readiness* and *Environment* – related capabilities.

## CONCLUSION

In this chapter, we developed and tested a novel framework, supported by a hybrid DEA/DT methodology, the purpose of which is assisting researchers and practitioners to investigate impacts of ICT capabilities on micro-economic outcomes. The process of testing of our methodology allowed us to answer three research questions, leading us to the answer to the general research question of this study:

> *What are the context-specific factors that differentiate various groups of economies in terms of the impact of ICT?*

However, as we demonstrated to our reader the differentiating factors are dependent on the meaning, or operational definition of the word *various*. This is because, just in our illustrative example, a group of countries could be subdivided into sub-groups based on a classification of IMF, or based on a relative efficiency of conversion of capabilities into outcomes, or based on growth in productivity. The subdivision could also be made on the basis of combining multiple criteria. For example, an investigator could ask "What differentiates relatively inefficient Sub-Saharan economies that exhibited growth in productivity from relatively efficient advanced economies whose productivity declined"? Furthermore, a subdivision could even be made on the basis of a given *ICT Capability → Microeconomic Outcome* path. Thus, for illustrative purposes, we answer the general question of our investigation at the very low level of granularity, as it applies to one arbitrary chosen case – we compare Ghana and Czech Republic with regards to *Usage → State of Labor* path. We present the results of the analysis in Table 20.10.

The results of the analysis offer evidence that the proposed framework could be used as a valuable tool benefitting a policy maker in the area of ICT4D.

**TABLE 20.10**
**Illustrative Example: Answer to the General Research Question of the Study**

| Criterion/Economy | Ghana | Czech Republic | Differentiating Factor |
|---|---|---|---|
| IMF Classification | Sub Saharan Africa | Advanced Economy | Czech Republic has a higher level of ICT capability in the area of *Market Environment* then Ghana |
| Relative Efficiency (*Usage → Labor*) | Inefficient | Efficient | Ghana has higher level of ICT capability in the area of *Business Readiness* than Czech Republic |
| Source of Growth, Productivity (*Usage → Labor*) | Technological Change | Change in Efficiency | For Ghana, *technological change* is a primary source of growth in productivity, while for Czech Republic the primary source is *change in efficiency*. |

## CONTRIBUTIONS TO THEORY

While the framework of neoclassical growth accounting allows investigators for inquiring in the macroeconomic impact of ICT, it does so at the very high level of granularity. Based on the Cobb-Douglas formulation we can only state that investments in ICT *do* contribute to the macroeconomic outcomes, while the *how* – the mechanism of the contribution is, in essence, a "black box". The proposed framework, while being consistent with the framework of neoclassical growth accounting, allows investigators to gain insights into possible mechanics of that "black box" and inquire into paths by which investments in ICT contribute to macroeconomic outcomes. As noted earlier benefits of the proposed framework and methodology include the enabling the identification of those ICT capabilities that are lagging within a given context; and enabling the identification of those mechanisms of transformation of ICT capabilities into macroeconomic outcomes that need to be adjusted. Consequently, we consider *an elaboration of the framework of neoclassical growth accounting* to be one of the theoretical contributions of this study.

Another contribution of this research to theory is represented by *the model of sustainable impact of ICT.* This model is consistent with the framework of neoclassical growth accounting and could be applied to any type of capability-producing investments. Finally, we have demonstrated a theoretically-sound process of framework building that relies on three steps – identifying capabilities, identifying impacts, and identifying mechanisms of transformation of capabilities into impacts. We consider this *domain-independent three-step process of framework building* to be another theoretical contribution of this research.

## CONTRIBUTIONS TO PRACTICE

The main practical contribution of this study is represented by *the context-specific application of the methodology*. We have demonstrated how DEA and DT could be applied to answer context-specific questions that are relevant to ICT4D. The actual results of the data analysis should also be considered as a contribution to practice, for they offer an additional perspective on the state of the economies in our sample. Finally, by incorporating Networked Readiness Index in our investigation we have been able to provide yet another dimension to the scores of NRI – we were able to illustrate the meaning of the scores in relation to various microeconomic impacts. Specifically, we were able to demonstrate some practical implications of NRI-based ranking of economies.

## ACKNOWLEDGMENT

Material in this chapter previously appeared in: An analytical framework for exploring context-specific micro-economic impacts of ICT capabilities. *Information Technology for Development*, *24*(4), 633–657.

## REFERENCES

Daveri, F. (2000). Is Growth an Information Technology Story in Europe too? EPRU Working Paper Series 00-12, Economic Policy Research Unit (EPRU), University of Copenhagen. Department of Economics.

Dutta, S., Bilbao-Osorio, B., and Gieger, T. (2012). The Networked Readiness Index 2012: Benchmarking ICT Progress and Impacts for the Next Decade, Global Information Technology Report 2012, pp. 3–34. World Economic Forum. Available on-line at: http://www.weforum.org/reports/global-information-technology-report-2012.

Dutta, S. and Jain, A. (2003). The Networked Readiness of Nations. In Dutta, S., A. Lanvin, B., & Paua, F. (eds.), The Global Information Technology Report 2002–2003, Oxford University Press, New York, NY, Oxford.

Dutta, S., & Mia, I. (2011). Global information technology report 2010–2011: Transformations 2.0. Geneva: World Economic Forum.

GITR. (2012). Global information technology report. Retrieved from http://www3.weforum.org/docs/ Global_IT_Report_2012.pdf

Greenhill, R. (2010). Preface, Global Information Technology Report 2009–2010, World Economic Forum. Available on-line at: https://www.itu.int/wsis/implementation/2010/forum/geneva/docs/publications/GITR%202009-2010_Full_Report_final.pdf.

Hernando, I., and Nunez, S. (2002). The Contribution of ICT to Economic Activity: A Growth Accounting Exercise with Spanish Firm-Level Data. Banco de España Working Papers 0203, Banco de España.

Jorgenson, D.W., and Stiroh, K.J. (2000). US economic growth in the new millennium. Brooking Papers on Economic Activity, 1, 125–211.

Nielsen, L. (2011). Classifications of Countries Based on their Level of Development: How it is Done and How it Could Be Done. IMF, Working Paper No. 11/31. Available online at: https://www.imf.org/en/Publications/WP/Issues/2016/12/31/Classifications-of-Countries-Basedon-their-Level-of-Development-How-it-is-Done-and-How-it-24628

Oliner, S.D. and Sichel, D.E. (2000). The resurgence of growth in the late 1990s: is information technology the story? Journal of Economic Perspectives, 14(4), 3–22.

Samoilenko, S. (2013). Investigating factors associated with the spillover effect of investments in telecoms: do some transition economies pay too much for too little? Journal of Information Technology for Development, 19(1), 40–61.

Samoilenko, S. (2014). Investigating the impact of investments in telecoms on microeconomic outcomes: conceptual framework and empirical investigation in the context of transition economies. Journal of Information Technology for Development, 20(3), 251–273.

Samoilenko, S. and Ngwenyama, O. (2011). Understanding the human capital dimension of ICT and economic growth: an approach to analyzing different ICT workforce and technology investments policies. Journal of Global Information Technology Management, 14(1), 59–79.

Samoilenko, S. and Osei-Bryson, K.M. (2008). An exploration of the effects of the interaction between ICT and labor force on economic growth in transitional economies. International Journal of Production Economics, 115, 471–481.

Samoilenko, S., and Osei-Bryson, K.M. (2016). Human development and macroeconomic returns within the context of investments in telecoms: an exploration of transition economies. Journal of Information Technology for Development, 22(4), 550–561 (DOI:10.1080/02681102.2013.859116).

Schreyer, P. (2000). The Contribution of Information and Communication Technology to Output Growth a Study of the G7 Countries. OECD Science, Technology and Industry Working Papers 2000/2, OECD, Directorate for Science, Technology and Industry.

Solow, R. (1957). Technical change and the aggregate production function. Review of Economics and Statistics, 39(3), 312–20.

Whelan, K. (2000). Computers, Obsolescence, and Productivity. Federal Reserve Board. Finance and Economics Discussion Series 2000-06, Board of Governors of the Federal Reserve System (U.S.).

# 21 A Methodology for Identifying Sources of Disparities in the Socio-Economic Outcomes of ICT Capabilities in SSAs

## INTRODUCTION

It is expected that the impact of information and communication technologies (ICTs) at the country level extends beyond purely economic gains (e.g., via growth in productivity) and into the sphere of social development (Eide, 2015). While wealthier economies of the world may look towards optimization of the economic impact of ICT, poorer countries of Sub-Saharan Africa (SSA) should be in a position of reaping a transformational-level of socio-economic benefits of ICT. It is hard to determine whether the transformational impact within the context of SSA is indeed taking place, but we could start the assessment by investigating the link between the state of ICT and its socio-economic impacts. The premise is that for a sustained transformational impact to take place an economy needs to obtain and maintain an efficient path of transforming ICT capabilities into socio-economic outcomes.

Benchmarking is one of the tools by which improvements in efficiency could be obtained, and we suggest that this tool could be utilized by SSA economies to improve the performance of their *ICT Capabilities*. However, it is a bridge too far for SSA economies to benchmark developed countries outright, for the disparity in the levels of accumulated and developed ICT infrastructure and annual investments is too great to disregard. Thus, we suggest that as a first step we investigate the efficiency of socio-economic impact of ICT capabilities within a group of SSA. Such investigation would entail identification of the economies that are more efficient in obtaining socio-economic benefits of ICT, and then proceeding with identifying the characteristics of such economies vis-à-vis characteristics of the less efficient economies, all within the context of SSA.

We feel that such inquiry is well-justified, because "…the complex relationships between ICTs and socioeconomic performance are not fully understood and their causality not fully established" (Di Battista, Dutta, Geiger, & Lanvin, 2015, p. 4).

Consequently, this problem presents an important research opportunity to investigate and we formulate the mail goal of this study as follows:

> *The purpose of this investigation is to identify some of the factors that differentiate groups of Sub-Saharan economies in regard to their levels of wealth and efficiency of socio-economic impact of ICT.*

This would require accomplishing the following two research objectives:

1. *To develop a methodology for identifying a set of group-specific characteristics of economies reflecting their state of economic development and efficiency of socio-economic impacts of ICT.*
2. *To apply the developed methodology within the context of Sub-Saharan economies to identify factors associated with group-specific disparities in economic development and socio-economic impact of ICT.*

Resultantly, by conducting this investigation we aim to contribute to the existing body of knowledge in the area of ICT4D in more than one way. First, we develop a methodology allowing for identifying a combination of characteristics describing various groups of SSA. While decision trees analysis could be performed to identify the factors specific to various groups of economies (Samoilenko & Osei-Bryson, 2014), this technique is not well suited for identifying *combinations* of factors. In this study we demonstrate how association rule mining (ARM) and decision tree induction (DTI) could be used *in synergy* to identify a set of attributes differentiating various groups of SSA economies. The true novelty of this study is that our approach allows for *identifying sets of attributes based on differentiating factors.* Second, while the previous inquiries concentrated either on economic or on social impacts of ICT capabilities, our study aims to be more comprehensive in this regard – we investigate both social and economic impact of ICT capabilities within the same sample of SSA economies. Finally, the results of empirical analysis should provide valuable information to policy and decision makers working in the area of ICT4D within the context of SSA.

To summarize, accomplishing our research objectives would contribute to the current state of knowledge in the area of ICT for development in several ways including:

- *A development of a novel three-phase methodology for identifying a set of rules and differentiating factors that, taken together, allow for gaining deeper insights in disparities between the groups of economies.*
- *An identification of the factors and complex associations impacting the disparity of the economic development, socio-economic impact of ICT, as well as of the efficiency of the impact in the context of SSA*

We conduct our investigation within the context of 27 economies of SSA, using the data set for the period of 2011–2014. The analysis of the data is supported by a three-phase methodology utilizing data envelopment analysis (DEA), DTI, and ARM.

## RESEARCH FRAMEWORK

In our investigation we rely on the framework of Networked Readiness (Dutta, Geiger, & Lanvin, 2015), the adapted version of which is depicted in Figure 21.1. The framework relies on four subindexes and their ten sub-categories (or *pillars*) to obtain the value of the Networked Readiness Index (NRI), which reflects the capacity of economies to benefit from ICT. An increase in the value of NRI for a given economy is indicative of the increase of the impact of ICT on innovation and productivity (Dutta & Jain, 2003). Interestingly, the original framework does not explicitly connect *Environment*, *Readiness*, and *Usage* subindexes (referred to as *Drivers* within the framework) with *Impact* subindex (referred to as *Impact*), despite relying on a principle that "…the environment, readiness, and use—interact, co-evolve, and reinforce each other to create greater impact" (Di Battista, Dutta, Geiger, & Lanvin, 2015, p.4).

We scope our inquiry by only considering relationships between *Drivers* (environment, readiness, and use) and *Impact* (socio-economic impact of *Drivers*) – as indicated by arrows in Figure 21.1. All possible interactions within *Drivers* (e.g., between environment, readiness, and use) we consider to be beyond the scope of our investigation. In our inquiry we use the framework depicted in Figure 21.1 to investigate, via DEA, the efficiency of the process by which *Environment*, *Readiness*, and *Usage* subindexes impact two subcategories of *Impact* subindex – *Economic* and *Social* impacts. Once we identify the better and worse performers of the sample, we utilize DTI to identify some of the factors that differentiate groups of economies in regard to their level of economic development and efficiency of socio-economic impact of ICT. Then we use ARM, via *Market Basket Analysis*, to attempt to identify a set of rules describing the groups.

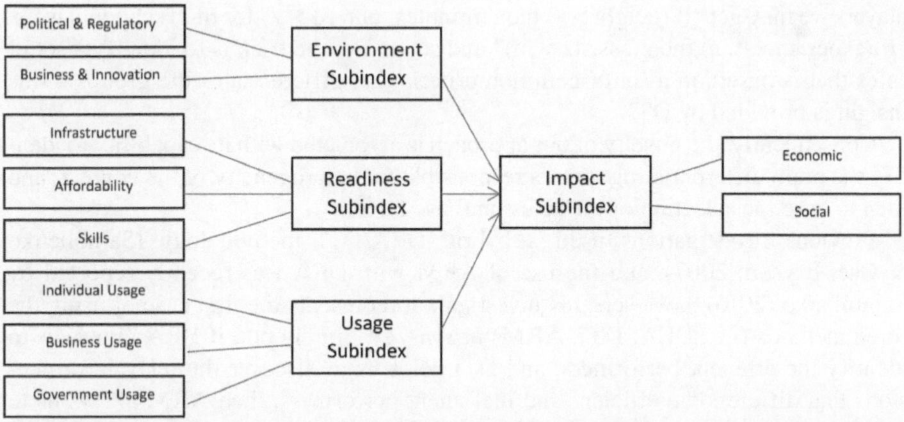


**FIGURE 21.1** The framework of networked readiness, adapted format.

## PROPOSED METHODOLOGY

Our proposed methodology consists of three phases where each phase involves the application of well-established data analysis methods (i.e., DEA, DTI, and ARM). Prior offering a description of the phases of our methodology we would like to explain to our reader what makes our approach truly novel.

### A New Methodology: Benefits and Justifications

DEA is a method that is widely used for the purposes of calculating scores of the relative efficiency of entities that receive inputs and produce outputs. For example, we could compare three groups of basketball players of different levels (e.g., high school, college, and professional) in terms of their efficiency of conversion of minutes played and attempts taken into assists and points. Results of DEA would yield the most relatively efficient group, but because DEA model is a "black box" model we would not know what differentiates the groups, or why one group is more efficient than the other two.

The insights could be provided by DTI, which would yield an attribute, or a few attributes, that differentiate the groups. Thus, given a set of attributes describing the three groups of basketball players we may find out that the main difference between the groups is in years of experience and hours of weekly practice. However, there are plenty of players with many years of experience who train long hours every week, but, nevertheless, don't play so well. We would like to know what set of attributes is actually associated with the outputs – assists or points scored. Here is where ARM may help.

ARM allows for generating sets of association rules of the type "if (a,b,c) then (d)". This is very valuable, for we can see the patterns of associations specific to each group. However, ARM tends to generate many rules, some trivial, some meaningless/non-actionable, and some useful. The problem with selecting the rules describing the different group is that the rules may contain completely different attributes – this would result in comparing apples and oranges. For example, in the case of basketball players we may get "if (height > x) then (minutes_played > y" for high school players, "if (experience > n) then (assists > m)" and so on. So, the trick is to identify a set of rules that is based on a set of common criteria that differentiates the groups – this insight is provided by DTI.

Consequently, the novelty of our approach is associated with its capability to identify the main differentiating factors responsible for heterogeneity of the context, and then to base the selection of the rules on those factors.

Previous investigations used a hybrid DEA/DTI methodology (Samoilenko & Osei-Bryson, 2007), and the use of ARM with DEA was recently reported by Samoilenko (2016); however, this investigation represents the first case of using the three methods (i.e., DEA, DTI, ARM) *in synergy*. Simply put, if DEA allows us to identify the efficient performers, and DTI helps us to discover the relevant dimensions that differentiate efficient and inefficient performers, then ARM allows us to benchmark efficient performers via a set of "IF THEN" rules that rely on the discovered by DTI dimensions. To our knowledge, no other combination of data analytic and data mining methods could offer so much in so few steps.

## Phase 1: Data Envelopment Analysis (DEA)

During the first phase we rely on DEA to evaluate relative efficiency of three "Drivers → Impact" paths. We will use *variable return to scale* (VRS) DEA model to conduct the analysis, for it is reasonable to argue that SSA economies have not yet reached the point of developing a level of ICT infrastructure allowing accruing the benefits yielded by capitalizing on economies of scale.

Given a four-year time period we will run DEA 12 times. Consequently, for each economy in the sample we are going to have four scores of relative efficiency for each of the three models. At this point we need to provide a justification for the inputs and outputs included in our models. In regard to outputs the reasoning is intuitive – first, we would like to assess the efficiency of the overall impact, and then, each type of the impact separately. This is because an economy could be efficient in obtaining one type of an impact (e.g., economic) and not efficient in regard to another impact (e.g., social) (Table 21.1).

With regard to the choice of the inputs of DEA model, our approach is methodological. While we are free to use eight sub-categories of Drivers as inputs of a DEA model, the general rule of thumb is that for a reasonable level of discrimination number of economies (or *Decision Making Units* in DEA terms) must be at least twice the product of inputs and outputs (Dyson, Allen, Camanho, Podinovski, Sarrico, & Shale, 2001). In our case we have a sufficient number of economies in our set, but if we use a DEA model with eight inputs and two outputs then we would need to have at least $2 \times 8 \times 2 = 32$ economies in the sample.

Furthermore, and more importantly, the greater the number of factors included in the DEA model, the lower the level of discrimination of the model (Dyson et al., 2001). However, we would like to use all the data available to us so we could inquire, for example, whether a set of specific factors – pillars – differentiate relatively efficient economies from relatively inefficient once. We would use DTI to do so.

Additionally, we use DEA to calculate the values of the Malmquist index (MI) – this allows us to assess the changes in relative efficiency of SSA economies that took place over time. Such results would not only identify the economies that exhibited growth

**TABLE 21.1**
**DEA Models of the Study**

| DEA Model | Inputs of DEA Model | Outputs of DEA Model |
|---|---|---|
| *Drivers* → Overall_*Impact* (DOI) | Environment Subindex<br>Readiness Subindex<br>Usage Subindex | Impact Sub-index |
| *Drivers* → Economic_*Impact* (DEI) | Environment Subindex<br>Readiness Subindex<br>Usage Subindex | Economic Impact Sub-category |
| *Drivers* → Social_*Impact* (DSI) | Environment Subindex<br>Readiness Subindex<br>Usage Subindex | Social Impact Sub-category |

in productivity (under assumption of constant return to scale), but to also identify the sources of growth (EC – change in efficiency vs. TC – change in technology). By applying DTI, we can identify factors that differentiate the *Growth vs. No Growth* economies.

### Phase 2: Decision Tree Induction (DTI)

To proceed with Phase 2 we need to create a new variable "Target" to differentiate various groups of economies. We are interested in three types of groupings: first, we would like to differentiate the groups of SSA by their level of economic development, and then we would like to differentiate the groups in terms of their efficiency of the socio-economic impact of ICT. The last analysis would involve differentiating SSA economies by growth in productivity – *Growth vs. No Growth.* Thus, we would conduct DTI three times, which would require Target to have three domains of values.

In the first case, grouping by income, the domain of values of *Target* would be {1, 2, 3}, for, respectively, Low Income (LI), Low Middle Income (LM), and Upper Middle (UM) groups of economies. In the second case, grouping by efficiency, Target would assume the values of {0, 1}, for, respectively, relatively inefficient, and relatively efficient SSA. The same domain of values, namely, {0, 1}, could be applied to the grouping by growth in productivity, where "0" would indicate "No Growth" and "1" would indicate "Growth".

### Phase 3: Association Rule Mining (ARM)

The purpose of Phase 3 is to find possible patterns, associations, or causal structures that may exist in our data. One of the main advantages of ARM is that it is suitable for undirected data mining; thus, we'll aim to discover naturally occurring associations between the factors (sub-indexes of *Drivers* and *Impact*) – components of NRI. ARM could be classified as either being explanatory or exploratory in nature. In the case of our investigation we employ exploratory ARM, for we do not have any theoretical support for why certain relationships between the sub-indexes of NRI should exist. A very common approach to generating associations between the variables, or *itemsets*, via ARM is by using the *a priori* algorithm (Agrawal & Ramakrishnan, 1994) – we will rely on this approach in the current investigation. Transformation of the data is required for this step – we follow the method of Samoilenko (2016) to do so.

## RESEARCH QUESTIONS AND NULL HYPOTHESES OF THE STUDY

At this point, we can operationalize the two objectives of this investigation in the form of the specific research questions and corresponding null hypotheses.

The first research question operationalizes the first objective as follows:

*Is the developed methodology capable of generating sets of differentiating factors and association rules for a given set of criteria?*

One of these criteria is associated with the level of income of economies (e.g., Low Income vs. Low-Middle vs. Upper-Middle), while another criterion is a relative level of efficiency (e.g., relatively efficient vs. relatively inefficient) of *Drivers → Impact* path, and the third one is a growth in productivity that took place over period of time.

We can answer this research question by testing the corresponding null hypotheses:

*H01a: The DTI part of the methodology will fail to generate a set of differentiating factors characterized by high-level splits.*

We will test H01a under the conditions of: *high-level splits that differentiate at least 60% of at least one of the groups of SSA economies.*

*H01b: The ARM part of the methodology will fail to generate sets of association rules for a given set of criteria.*

We will test H01b under the minimal conditions of *Support* > 20%, *Confidence* > 1.0, and *Lift* > 1.0.

While the results of DTI and ARM may offer useful insights by themselves, we would like to use the two methods in a complementary fashion; thus, we state another hypothesis as follows:

*H01c: The results of DTI and ARM are not complementary.*

We will test H01c under the condition that the differentiating factors identified by DTI would be included in sets of rules identified by ARM.

The second research question operationalizes the second objective as follows:

*Does the choice of criteria such as level of economic development, relative efficiency, and growth in productivity impact the combination of factors describing various groups of economies and relationships between Drivers and Impacts of ICT?*

Basically, we would like to find out if the different criteria could be characterized via different set of factors – this allows us to inquire into the specificity of a setting expressed as a combination of sub-categories of NRI. We will answer the second research question after testing our second null hypothesis:

*H02: No combination of factors contained in the generated association rules would be unique to a given context.*

The simple side-by-side comparison of the generated association rules and split variables will serve as a sufficient criterion for testing H02.

## THE DATA

We obtained the data from a reputable source – the World Economic Forum's Global Information Technology Report 2015 (GITR, 2015). In 2012 the representation of NRI was partially changed in terms of the number and representation of the pillars

**TABLE 21.2**
**Sample of Sub-Saharan Economies, by Income Level**

| Income Level | Sub-Saharan Economies |
|---|---|
| Low Income | Burkina Faso, Burundi, Chad, Ethiopia, Gambia, Kenya, Madagascar, Malawi, Mali, Mozambique, Rwanda, Tanzania, Uganda, Zimbabwe |
| Low-Middle Income | Cameroon, Cape Verde, Côte d'Ivoire, Ghana, Lesotho, Nigeria, Senegal, Swaziland, Zambia |
| Upper-Middle Income | Botswana, Mauritius, Namibia, South Africa |

of three sub-indexes of NRI; it was also the year when the *Impact* subindex was introduced. Given the changes that took place between 2011 and 2012, we decided to concentrate on the new version of NRI and collect the data provided in GITR 2012, 2013, 2014, and 2015. In some cases, the representation of SSA economies was inconsistent – for example, we could not include Angola, Seychelles, Liberia, Gabon, Sierra Leone, and Guinea in our sample because the data for some of the years was missing.

While there is an advantage to increasing the sample size of a study, there is a price to pay via dealing with missing variables, imputation of values, and additional data preprocessing. After considering the pluses and minuses of "sample size vs. data *actually* available" we have assembled a smaller data set that contained no missing data and no outliers, but was as reliable as one could get from a given source. Overall, we were able to compile the data set representing 27 economies of SSA (the classification of the International Monetary Fund as of October 2014). The sample consists of 14 low income economies, nine low-middle economies, and four upper-middle economies (the classification of the World Bank as of July 2014). Membership of each group of the sample is provided in Table 21.2.

## RESULTS OF THE DATA ANALYSIS

### Phase 1: Data Envelopment Analysis

We offer a summary of the results of DEA below. If a given economy has been determined to be relatively efficient for at least three times over the period of four years, we have labeled such economy as "efficient" for the whole period of four years. Because our economies fall within three distinct groups – low income (LI), low-middle income (LM), and upper-middle income (UM), we also determined the relative efficiency of each economy over the 4 years within its group – we will use this information in Phase 3 when we perform ARM.

Our results demonstrated that seven economies out of the full sample are relatively efficient with regard to the impact of *Drivers* on social, economic, and overall *Impact* of ICT. Additionally, we identified relatively efficient economies per each of the income-level group; in some cases (e.g., Burundi, Chad, Kenya, Mali, Rwanda, and Senegal), the relatively efficient within its group' economies are also efficient overall. In other cases (e.g., Swaziland, Lesotho, Botswana, Mauritius, Namibia, and

**TABLE 21.3**
**Results of DEA**

| Income Level | Economy | Overall Efficiency of the Impact of ICT | Overall Changes in Productivity | Growth via EC? | Growth via TC? |
|---|---|---|---|---|---|
| LI | BFA | Efficient | Growth | Yes | Yes |
| | BDI | Efficient | Growth | Yes | No |
| | TCD | Efficient | No growth | No | No |
| | ETH | Efficient | No growth | No | No |
| | GMB | Inefficient | Growth | No | No |
| | KEN | Efficient | No growth | No | No |
| | MDG | Inefficient | Growth | Yes | No |
| | MWI | Inefficient | No growth | No | No |
| | MLI | Efficient | Growth | No | Yes |
| | MOZ | Inefficient | No growth | Yes | No |
| | RWA | Efficient | No growth | Yes | No |
| | TZA | Inefficient | Growth | Yes | No |
| | UGA | Inefficient | No growth | Yes | No |
| | ZWE | Inefficient | Growth | Yes | No |
| LM | CMR | Inefficient | Growth | Yes | No |
| | CPV | Inefficient | No growth | No | No |
| | GHA | Inefficient | No growth | No | No |
| | NGA | Inefficient | No growth | No | No |
| | SEN | Efficient | No growth | No | Yes |
| | SWZ | Inefficient | Growth | Yes | No |
| | ZMB | Inefficient | Growth | Yes | No |
| | CIV | Inefficient | Growth | Yes | No |
| | LSO | Inefficient | Growth | Yes | No |
| UM | BWA | Inefficient | No growth | No | No |
| | MUS | Inefficient | Growth | Yes | No |
| | NAM | Inefficient | No growth | Yes | No |
| | ZAF | Inefficient | No growth | No | No |

South African Republic), the relatively inefficient, overall, economies end up being efficient within their respective group (Table 21.3).

Additionally, we used DEA to calculate the values of Malmquist Index (*MI*), which allows us to assess the changes in the scores of relative efficiency that took over period of time. Under the assumption of constant returns to scale the change indicates changes in productivity. Consequently, we are able to assess whether the economies become more productive or not. Because *MI* is comprised of two components – change in efficiency (*EC*) and change in technology (*TC*), we are also able to assess whether the changes in productivity are associated with a particular component. Overall, only 11 economies (40% of the sample) exhibited growth in productivity.

An analysis of the changes in *EC* and *TC* offers an interesting insight: 16 economies (60% of the sample) exhibited positive changes in efficiency, but only 3 economies (11% of the sample) demonstrated positive changes in technology. Finally, it is worth noting that only one economy, Burkina Faso, exhibited a balanced growth in productivity, when the growth was driven by both components of *MI*. Overall, the picture suggests that SSA, as a group, would benefit from a better technology – this suggests that investments in ITC infrastructure should be prioritized. Finally, it is worth noting that eight economies (30% of the sample) have not only exhibited a decline in productivity, but have exhibited a decline in terms of both components of *MI*.

Overall, we summarize the results of DEA part of our methodology as follows:

- Seven economies out of the full sample are relatively efficient with regard to the impact of *Drivers* on social, economic, and overall *Impact* of ICT
- There are relatively efficient economies per each of the income-level group
- Only 11 economies (40% of the sample) exhibited growth in productivity, while 8 economies (30% of the sample) have exhibited a decline
- SSA, as a group, would benefit from a better technology.

## Phase 2: Decision Tree Induction

The results of DTI allow us to test our first null hypothesis, H01a, for decision tree induction did generate high-level splits that differentiated groups of economies. Results summarized in Table 21.4 show that such pillars of NRI as *Individual Usage*, *Business Usage*, and *Skills Readiness* do play important role in differentiating three groups of economies. It is not surprising that there is appear to be a clear-cut difference between Low Income and Upper-Middle Income economies, and much less of a difference between Low-Middle Income economies and the other two.

We could also identify *Individual Usage* and *Economic Impact* as pillars that play role in differentiating relatively efficient SSA economies from inefficient ones. It appears that Infrastructure Readiness, Affordability Readiness, and Individual

**TABLE 21.4**
**Results of Decision Trees Analysis**

| Grouping by | Group | Differentiating/Split Variable |
|---|---|---|
| Economic development | Low Income vs. Low Middle Income vs. Upper Middle Income | Individual Usage<br>Business Usage<br>Skills Readiness |
| Relative efficiency | Relatively Efficient vs. Relatively Inefficient | Individual Usage<br>Economic Impact |
| Change in productivity | Growth vs. No Growth | Infrastructure Readiness<br>Affordability Readiness<br>Individual Usage<br>Social Impact |

Usage are factors playing role in differentiating those economies that became more productive from those that didn't.

At this point, we summarize the results of DTI part of our methodology as follows:

- Pillars of NRI such as *Individual Usage*, *Business Usage*, and *Skills Readiness* do play important role in differentiating three groups of economies
- *Individual Usage* and *Economic Impact* are pillars that play a role in differentiating relatively efficient SSA economies from inefficient ones
- Infrastructure Readiness, Affordability Readiness, and Individual Usage differentiate those economies that became more productive from those that didn't.

## Phase 3: Association Rule Mining

The results summarized in Table 21.5 allow us to test our null hypotheses. First, the results allow us to reject *H01b*, for the application of ARM did result in the generation of multiple association rules under the criteria of *Support* > 20%, *Confidence* > 1.0, and *Lift* > 1.0. Second, the results also allow us to reject *H02*, for the generated by ARM rules contain context-specific combinations of factors.

**TABLE 21.5**
**Impact-Specific Rules for Low Income and Low-Middle Income economies**

| Condition | Generated Rules | | | Sup. | Conf. | Lift |
|---|---|---|---|---|---|---|
| LI: Low Income | low IND_USE, low BUS_USE | → | low ECON_IMP | 21% | 0.7 | 1.9 |
| | low SKILL_READ, low BUS_USE | → | low ECON_IMP | 21% | 0.6 | 1.5 |
| LM: Low-Middle | midhigh BUS_USE | → | midlow SOCIO_IMP | 20% | 0.5 | 2.5 |
| | midhigh SKILL_READ | → | midhigh SOCIO_IMP | 25% | 0.5 | 1.5 |
| UM: Upper-Middle | high SKIL_READ, high AFFORD_READ, high GOV_USE | → | high SOCIO_IMP | 32% | 1.0 | 2.3 |
| | high BUS&INNOV_ENV, high INFR_READ, high BUS_USE | → | high ECON_IMP | 31% | 0.85 | 2.3 |
| | high BUS&INNOV_ENV, high IND_USE, high BUS_USE | → | high ECON_IMP | 31% | 0.85 | 2.2 |
| Low Income, Inefficient | low INFR_READ, low BUS_USE | → | low SKILL_READ | 36% | 1.0 | 1.9 |
| | low IND_USE, low BUS_USE | → | low SKILL_READ | 32% | 1.0 | 1.9 |
| Low-Middle, Efficient | low BUS_USE, low GOV_USE | → | low ECON_IMP | 25% | 1.0 | 4.0 |
| | low BUS_USE, low ECON_IMP | → | low SOCIO_IMP | 25% | 1.0 | 3.0 |

Finally, the results summarized in Table 21.5 allow us to reject *H01c* that *the results of DTI and ARM are not complementary. DTI identified Individual Usage, Business Usage*, and *Skills Readiness as the variables differentiating the groups of economies in our sample.*

## DISCUSSION OF THE RESULTS

The results of the data analysis, presented in the previous sections, offer evidence that we were successful in addressing the research questions of this study. We developed and tested a novel methodology allowing for investigating complex context-specific relationships between the factors reflecting the state and the impact of ICT capabilities. The discussion of the results is presented along the points that we considered noteworthy.

First,

> *Despite the presence of complex relationships between the Drivers and Impacts of ICT there are common themes associated with the levels of the scores of factors comprising NRI – Business Usage, Individual Usage, and Skills Readiness appear to have a direct relationship with the levels of the scores of socio-economic Impact of ICT.*

We point out that while the variety of association rules has been generated for a different set of criteria, a common line could also be glanced – some subcategories of NRI' subindexes (e.g., related to Skills, Business, Individual usage) appear more frequently than other subcategories.

Second,

> *Results of our investigation suggest that Business Usage and Individual Usage are among the factors that appear to differentiate economies in terms of their level of economic development, as well as in terms of their relative efficiency of the impact of ICT on the socioeconomic bottom line.*

These results suggest that wealthier and more efficient economies tend to have higher scores of *Business Usage and Individual Usage*. The presence of a simple association between the level of income of an economy, its efficiency, and ICT usage seems to be apparent.

Third,

> *Infrastructure and Affordability of ICT seem to have an impact on growth in productivity of SSA economies.*

This finding is important because it was provided by two different methods of analysis – DEA and DTI. According to the results of DEA only three economies exhibited growth in technology over the period 2012–2015, and DTI independently confirmed it by identifying *Infrastructure* and *Affordability* as factors differentiating growth vs. no growth economies of SSA.

## CONCLUSION

In this investigation, we developed and applied a methodology allowing for generating sets of association rules from the combination of factors describing relationships between *Drivers* and *Impact* of ICT. The results of the data analysis do confirm the notion that the relationships between the factors representing *Drivers* and *Impact* are indeed complex. However, the underlying complexity of the relationship could be made more transparent to researchers and practitioners by the developed in this study methodology.

## ACKNOWLEDGMENT

Material in this chapter previously appeared in: A methodology for identifying sources of disparities in the socio-economic impacts of ICT capabilities in Sub-Saharan economies. In *International Conference on Information Resources Management (CONF-IRM)*. Association for Information Systems.

## REFERENCES

Agrawal, R., and Ramakrishnan, S. (1994). Fast Algorithms for Mining Association Rules in Large Databases. In Proceedings of the 20th International Conference on Very Large Data Bases (VLDB '94), Jorge B. Bocca, Matthias Jarke, and Carlo Zaniolo (Eds.). Morgan Kaufmann Publishers Inc., San Francisco, CA, pp. 487–499.

Di Battista, A., Dutta, S., Geiger, T., and Lanvin, B. (2015). The Networked Readiness Index 2015: Taking the Pulse of the ICT Revolution, Global Information Technology Report 2015, pp. 3–28.

Dutta, S., and Jain, A. (2003). The Networked Readiness of Nations. In The Global Information Technology Report 2002–2003, Dutta, S., A. Lanvin, B., & Paua, F. (eds.). Oxford University Press, New York, NY, Oxford.

Dutta, S., Geiger, T., and Lanvin, B. (Eds.) (2015). Global Information Technology Report 2015. ICTs for Inclusive Growth. World Economic Forum and INSEAD, Geneva. Retrieved April 20, 2015 from: http://www3.weforum.org/docs/WEF_Global_IT_Report_2015.pdf.

Dyson, R.G., Allen R., Camanho A.S., Podinovski, V.V., Sarrico C.S., and Shale, E.A. (2001). Pitfalls and Protocols in DEA. European Journal of Operational Research, 132, 245–259.

Eide, E.B. (2015). Preface. In Global Information Technology Report 2015, p. v.

GITR. (2015). World Economic Forum' Global Information Technology Report. Available on-line at: http://reports.weforum.org/global-information-technology-report-2015/network-readiness-index/.

Samoilenko, S. (2016). Disparity of Social and Economic Impact of ICT Capabilities in Sub-Saharan Economies: Empirical Investigation of Differentiating Factors, in Proceedings of the SIG GlobDev Pre-ECIS Workshop ICT in Global Development; Istanbul, Turkey, June 12, 2016.

Samoilenko, S., and Osei-Bryson, K.M. (2007). Increasing the Discriminatory Power of DEA in the Presence of the Sample Heterogeneity with Cluster Analysis and Decision Trees, Expert Systems with Applications, 34, 2, 1568–1581.

Samoilenko, S. and Osei-Bryson, K.M. (2014). Formulation of Context-Dependent and Target-Specific Strategies of the Impacts of ICT on Development. In Proceedings of the 7th Annual SIG GlobDev Pre-ICIS Workshop ICT in Global Development, Auckland, New Zealand, December 14, 2014.

# 22 Discovering Common Causal Structures that Describe Context-Diverse Heterogeneous Groups

## INTRODUCTION

It is commonly acknowledged that operational excellence is one of the sources of competitive advantage of modern enterprises. Fundamentally, the concept refers to achieving a high level of efficiency of conversion of inputs into outputs, where a higher level of efficiency implies, *ceteris paribus*, a greater degree of excellence. Modern organizations typically regard information and communication technologies (ICTs) as one of the significant direct or indirect inputs for achieving such operational excellence and competitive advantage. Since the concept of competitive advantage involves a relative comparison of the performance of organizational entities, the concepts of organizational capabilities and benchmarking (e.g., Gouveia, Dias, Antunes, Boucinha & Inácio, 2015) are relevant. Ayabakan, Bardhan & Zheng (2017) noted that with regards to the investigation of this pair of concepts: "A dominant approach in IS research involves the use of survey instruments designed to elicit user responses on their perceptions about competencies and capabilities … A limitation of such perception-based approaches is that they represent a subjective measure of firm/organizational capabilities"; these researchers therefore proposed an approach that involves the use of non-subjective data and the data envelopment analysis (DEA) method for doing benchmarking (e.g., Adler, Liebert & Yazhemsky, 2013; LaPlante & Paradi, 2015) of the *Input–Output* conversion process. In this chapter, we not only take a similar approach to benchmarking but are also interested in the context of the *Input–Output* conversion process, including those that involve ICTs as input(s). The motivating idea for this research project is that the process of benchmarking could possibly be enhanced by the discovery and application of non-obvious common causal structures that differentiate more efficient organizational entities from less efficient ones. This motivating idea triggered our intention to design an appropriate methodology artifact that involves the analysis of non-subjective data. This research project can be considered to fall within the realm of information systems (IS) research for at least the following reasons: (1) benchmarking research is an aspect of well-established IS/ICT & Productivity research stream (e.g., Hitt & Brynjolfsson, 1996; Ko & Osei-Bryson, 2004); (2) benchmarking research has appeared in leading IS journals (e.g., Ayabakan, Bardhan & Zheng, 2017); (3) the proposed solution artifact involves the a creative integration of several IS artifacts (i.e., multiple data mining methods) with DEA; and (4) our illustrative example falls within the well-established IS/ICT & Productivity research stream.

## A CONCEPTUALIZATION OF THE BENCHMARKING PROBLEM

The term "benchmarking" is a commonly used one, and popular terms tend to be vulnerable to falling prey to the unfortunate assumption of the universality of their meaning. The concept of benchmarking is important to our inquiry; thus, we feel it is warranted if we spend a few sentences making sure that we clarify the chosen meaning of the term to our readers. Fundamentally and historically, benchmarking means *accurate application of a measure*, whatever the measure of interest could be (Merriam-Webster, 2017). Consequently, benchmarking is inherently a two-part process. Firstly, the presence of some sort of a standard measure, or a *target*, is established. And secondly, the process of emulation of that target is undertaken. For example, when we use a ruler to draw a 10-inch line on a piece of a chapter, we are benchmarking a chosen target (e.g., 10 inches) in the context of the piece of a chapter.

We can generalize benchmarking as *a process of emulation of the target in a new context*. This, immediately, brings up the problem of representation of the target, for in order to emulate a target we need to know the attributes that sufficiently describe (e.g., represent) the target. This is not an easy undertaking in the case of complex targets. One thing is for a freshman to benchmark her academic performance by aiming to have 4.0 GPA (a trivial one-dimensional construct), and another thing for her to benchmark against the likes of Einstein (a non-trivial multi-dimensional construct). Similarly, it is easy for a competitor to benchmark the battery life of cellular phone of the industry leader, and it is hard to benchmark the phone itself.

The whole concept of formal representation, which is relevant to benchmarking specifically and underlies the whole field of computing generally (e.g., if it can be formally represented, then it can be computed), is dependent on two factors. One is *objectivity* and the other is *scope*. The factor of objectivity of representation refers to having an objective (e.g., standard, agreed upon) scale for a given characteristic of interest, where an attribute "Sugar Content" could be objectively represented via "grams per kilogram" scale, and subjectively so via scale "perceived sweetness". The factor of objectivity, which is dealt with by finding or creating an appropriate measurement scale, is much easier to address than the factor of scope. In simple terms, scope deals with selecting what is "in" and what is "out", deciding on a set of attributes that adequately model (e.g., describe, represent) the target. The complexity of the decision is directly related to the complexity of the target; consequently, it is easier to benchmark a body mass index (BMI) than a luxury car, or a successful firm.

This issue of scope is not a trivial one because from a complex systems perspective we don't have a philosophical basis for making the decision regarding what the adequate description of the system itself would be. While a component-based description seems to be good and easy beginning things get increasingly complicated once we start considering non-linearity of the relationships between the components, various dependencies, and emergent properties. Furthermore, in the case of complex adaptive systems (e.g., person, organization, or economy) we cannot completely "abstract away" the system's environment, which further complicates the undertaking of nice and neat scoping of the target. Under such circumstances, *the scope of the target of benchmarking is not given, but is a result of an active discovery*, and it is in this area that this chapter aims to make a contribution.

While the concept of benchmarking can be applied to a great variety of contexts, in this chapter we apply it to *economic units*, which we define as a set of entities (e.g., firm, economy) that: (a) transform a set of expense-associated inputs into a set of revenue-associated outputs and (b) aim to minimize expenses and maximize revenues.

Any viable economic entity aims to ensure its survival by means of adopting and maintaining a valid business model. One of the purposes of a business model is to ensure that the stream of revenues is greater than the stream of expenses. While a valid business model assures an operational-level day-to-day viability of the entity, it does not guarantee long-term survival, for an entity could be doomed due to failed strategic- and tactical-level initiatives, such as poorly chosen and implemented strategy, erroneous vision, or misguided business goals.

In order for a business model to bring the intended results, it must be implemented – this is done via business processes. For example, a business model "sale of product to customers" in the context of a bakery can be implemented by means of "acquire inventory", "bake products", and "sell baked goods" processes. It is an effective and efficient execution of business processes that constitutes the operational excellence of firms, and it is not surprising, therefore, that the less successful firms often aim to improve their operational performance by means of benchmarking of the business processes of their more successful counterparts.

Our focus in this research is on benchmarking of business processes within the context of economic units, and, conceptually, the problem that we are trying to help addressing is associated with the necessarily different contexts of the target and the destination of the benchmarking. Let us consider two bakeries – one being a target of benchmarking (e.g., highly efficient) and another one being a destination of benchmarking (e.g., less efficient). Clearly, the process of baking is important and is easy to define based on the "ingredients → baked products" model. However, the same process takes place within different contexts: two bakeries may have a different number of employees with a different number of years of experience, they may have different equipment, and they may have different environmental conditions. Thus, the context of the process of baking is also important but is difficult to define, for there is no common context model for the target and the destination.

Conceptually, the question can be expressed as:

*How to scope the target of benchmarking so it takes into consideration contextual factors that are relevant and applicable to the destination of the benchmarking?*

## RESEARCH PROBLEM AND RESEARCH QUESTIONS OF THE STUDY

Fundamentally, any business process can be seen as a process of conversion of means into ends, where the primary goal is to minimize the cost of means and maximize the value of ends via increasing efficiency and effectiveness of the mechanism of transformation. If we consider a concept of a business process from a structural perspective, we can identify three distinctive parts. First, there is a set of inputs, second, there is a set of outputs, and third, there is a mechanism of transformation of inputs into outputs. This structural decomposition of a business process is important because it allows identifying a component, or components, that are relevant to the process of benchmarking.

A simplified model of conversion of inputs into outputs will take into consideration only those inputs that are necessary for producing the required output. Such a "pure" *Input–Output* process is akin to a recipe, which specifies a set of ingredients required to make a dish. Once an optimal recipe is developed and acknowledged as the model to be emulated, be it for making pies, or fragrances, or cars, or cellular phones, it is becoming fixed and pretty much ready for benchmarking. There are, of course, well-known exceptions of "secret recipes" of Coca-Cola, Pepsi, KFC, Bush's, and McDonald, but those are what they are – exceptions to the rule of what otherwise can be called *recipe transparency.*

Let us consider an example of Company A trying to benchmark the process of production of Apple's iPhone 8. The output of the process is known – it is the phone itself. The inputs are also known – the teardown (e.g., Techinsights.com, 2018) provides a complete list of the inputs – required components. Furthermore, the process of conversion of inputs into outputs is also known – there is a specific way in which parts are assembled into the whole. Would Company A, knowing all the inputs, knowing the mechanism of transformation (assembly), and knowing the output, succeed in getting the same result as Apple have? The answer is simple: it depends on the context – in a perfect world, yes, and in a real world, no. It is the context that presents most of the problems, and the question is how to account for the differences in the contexts. We suggest that it can be done by means of expanding a set of inputs and associated outputs beyond what is required by the pure production process and incorporating relevant context-reflecting factors.

In addition to context-specific factors, a more realistic process of conversion of means into ends would also take into consideration inputs and outputs of a conceptual nature represented via proxies (e.g., "level of customer satisfaction" represented via "# of repeat purchases"). This is especially true for enterprises that must consider not only pure inputs and outputs of a production process but also a specificity of the context within which inputs are obtained and utilized, as well as the context for which the outputs are produced – outcomes of the process of conversion of inputs and outputs. Granted, a representation of a pure, chemistry lab-like, input–output process of conversion may serve well in push-dominated industries. However, one of the outcomes of globalization is a more prevalent pull-dominated marketplace that forces consideration of the context – where the ease of access to inputs and desirability of outputs are the intuitive factors that come to mind. An enterprise that makes bricks and an enterprise that specializes in fast fashion may both rely on operational excellence, but the two will differ greatly in regard to what, conceptually, constitutes a set of inputs and a set of outputs of their respective operations.

Even enterprises of the same type within the same industry (and even if owned by the same company (Korotin, Popov, Tolokonsky, Islamov & Ulchenkov, 2017) within the same industry) often require consideration of the context within which their operations take place. For example, a meaningful comparison of the efficiency of an oil rig in Siberia with an oil rig in Saudi Arabia must take into consideration the contexts within which rigs operate. Similarly, a measure of efficiency of a delivery company should take customer satisfaction into consideration by using some sort of a proxy on the output side of the conversion equation, for an "amount of fuel → distance travelled" model is of limited use. Consequently, firms must be looking for a model of the type of "Input→Output→Outcome", which, in the case of a delivery company would be expressed as "Amount of Fuel→Distance Travelled →Customer Satisfaction". In both cases, one would have to consider variables that are not endemic to the *Input–Output* process, but rather reflective of the environment within which the process takes place. Keeping this in mind we advance the following proposition:

> *An assessment of the level of operational excellence of an enterprise operating in pull-driven global marketplace must take into consideration the specificity of the context of the conversion of Inputs into Outputs as well as a measure of outcome of the process.*

Once we start considering the relevance of the context and outcomes of the process of *Input–Output* conversion to the idea of operational excellence we arrive at two intuitive implications. First, the possible domain of *Input–Output* attributes expands beyond the actual, "true" variables involved in the process. While a compounding pharmacy may rely on a "true" process of conversion of the "true" *Inputs* (i.e., chemical components), into "true" *Outputs* (i.e., drugs), an outlet mall will have a fundamentally different perspective on what the "true" process is, and what are the *Inputs* and *Outputs* of the process are.

Second, commonly utilized process management and process improvement techniques (e.g., Kaizen, Six Sigma, etc.) become insufficient due to their intra-process and output-oriented foci. One of the options to improve the level of operational excellence is by benchmarking (adoption of Enterprise Systems falls under this category), where a company seeks a worthy-to-emulate counterpart (within, or outside of one's industry). Another option is to re-engineer the existing process (e.g., via Business Process Re-engineering). Both approaches have been utilized, and both approaches have been criticized (Al-Mashari et al., 2001; Fui-Hoon Nah et al., 2001; Cao, Clarke, and Lehaney, 2001; Moon, 2008). In the case of Business Process Re-engineering (BPR), there is too much emphasis on the context with the goal of creating unique processes, and in the case of benchmarking via Enterprise Systems (ES), there is too little emphasis on the context of the existing business processes due to their replacement by the inherent processes embedded in the ES.

We suggest that the process of benchmarking could be enhanced by discovering non-obvious common causal structures that may be shared by the involved entities. Let us consider a distribution company that operates multiple warehouses in different states in the US. And let us suppose that we are interested in the process that we will call "Receiving a Shipment from a Manufacturer → Delivering the Product to the Department Store". Fundamentally, the process of increasing a level of efficiency relies on two insights: first, the insight regarding the process itself – the way the shipment is received, then stored, then accessed, and then shipped out. This is a straightforward optimization problem that could be tackled by century-old methods of Frederick Taylor (1914). The second insight, however, is more important, for it deals with the context of the process – what are the specific conditions that impact the process, and what are the conditions that differentiate less efficient warehouses from more efficient ones. Some of the conditions are uncontrollable – a crew running an oil rig in Siberia cannot do much about an insight "outside temperature of greater than 85 degrees → higher oil production" received from their counterpart in Saudi Arabia. However, some of the insights could be utilized – an insight "average experience of a crew member is greater than 5 years → higher oil production" is an actionable one.

This leads us to a second major proposition:

*The impact of the process of benchmarking that is geared towards increasing the level of operational excellence could be enhanced if it is augmented by the actionable insights, in the form of non-obvious causal structures, regarding the context within which the operations take place.*

At this point, we would like to illustrate the meaning of the concept of "non-obvious causal structures" to our reader. First, we define the concept of "causal structure" as any type of "if A, then B". An "obvious causal structure", accordingly, is a recipe-like depiction of a production process, exemplified by an equation for a commonly utilized concrete mix: "4 parts crushed rock + 2 parts sand + 1 part cement → 7 parts concrete mix".

Let us consider Table 22.1. The obvious meta-causal structure for both stores is in the form "Investments → Revenues", where if Store A operates under the model "High_Investments → High_Revenues", then Store B operates under the model "Low_Investments → Low_Revenues". The simple solution for Store B is to change the model to "High_Investments" and hope that "High_Revenues" will transpire on

**TABLE 22.1**
**Contextual Translation of Meta-Causal Structures into Specific Structures**

| Context | Meta Input | Specific Input | Specific Output | Meta Output |
|---|---|---|---|---|
| Generic Model | Investments, in $ | Inp_Target 1<br>Inp_Target 2, etc. | Out_Target 1,<br>Out_Target 2, etc. | Revenues, in $ |
| Store A: High operational efficiency,<br>High-profit ratio | High Level of Investments | Hiring more staff,<br>Staff Training,<br>Store Maintenance,<br>InventoryType_1,<br>InventoryType_2, etc. | Number of Associates,<br>Customer Satisfaction,<br>Sales_InventoryType_1,<br>Sales_InventoryType_2, etc. | High Level of Revenues |
| Store B: Low operational efficiency,<br>Low profit ratio | Low Level of Investments | Hiring more staff,<br>Staff Training,<br>Store Maintenance,<br>InventoryType_1,<br>InventoryType_2, etc. | Number of Associates,<br>Customer Satisfaction,<br>Sales_InventoryType_1,<br>Sales_InventoryType_2, etc. | Low Level of Revenues |

the right side. However, investments must be allocated toward something (see *Specific Input* column) – the question is how to allocate the money in the most beneficial way. If Store B benchmarks Store A, then certain causal structures may not be applicable. For example, due to the difference in climate "InventoryType_1 → Sales_InventoryType_1" model may work wonderfully well for Store A, but would not be applicable to Store B. An obvious causal structure "Hiring more staff → Number of Associates" will always work, but it is not clear if this is the best bang for a buck for Store B.

A set of possible non-obvious causal structures could be depicted by the model "Hiring more staff, Staff Training, Store Maintenance → Customer Satisfaction". Given limited resources of Store B, then to which one of the target inputs should investments be allocated? The question is simple to answer if we discover that Store A and Store B do not differ greatly in terms of number of associates, or money spent on store maintenance, but do differ significantly in terms of the money spent on staff training. Consequently, under this scenario a non-obvious causal structure would be "Staff Training → Customer Satisfaction", and if this model is currently implemented as "Staff Training= High → Customer Satisfaction=High" in Store A, and as "Staff Training=Low → Customer Satisfaction=Low", then Store B could consider investing more money in this area knowing that it worked in Store A.

Let us consider yet another example of a company that operates a set of similar retail stores in various locations in the US. At the very high level, the process that is shared by all the stores is "Investments in the store → Revenues from the store", but this process takes place in different contexts. In such cases, the term "context" means not only geographic location, climate, average income of the local population, etc. but also includes details regarding the application of investments. Meaning, when a certain amount of money is allocated, it is eventually allocated toward something – renovation of the store, upgrade of the equipment, and training of the staff. Furthermore, investments could be allocated toward purchasing different types of inventory – in some cases, it could be sporting goods, in other cases, it could be jewelry, or it could be household items. A less efficient store in Arizona may not benefit much from the insight "Increase in Inventory of Down Jackets → Increase in Revenue" offered by a more efficient store in Alaska, but it could benefit from an insight "hours of annual staff training >20 → increase in revenue". The trick then becomes is how to uncover the actionable insights that allow for increasing the level of operational excellence.

All the points made above could be summarized in the following set of propositions:

1. Less efficient enterprises could improve their performance via benchmarking their more efficient counterparts (a Minor League baseball pitcher could benchmark a Major League pitcher)
2. The difference in the level of efficiency is due to two factors – efficiency of the process of conversion of *Inputs* into *Outputs*, and the context within which the process takes place (a Major League pitcher could be better than a Minor league pitcher because of the superior technique – the process of pitching, or due to the context within which the process is utilized – the experience, strength, height, hours spent in the gym, etc.)

3. Some of the contextual factors that impact the process are shared by less and more efficient enterprises are controllable – by improving these factors less efficient enterprises can increase the level of efficiency of the *Input–Output Process* (while a Minor League pitcher cannot become taller, he can increase his level of efficiency by spending more time in the gym)
4. Discovering a set of controllable contextual factors would enhance the result of a benchmarking, thus allowing less efficient enterprises to increase their level of efficiency.

Taking into consideration these examples and propositions, at this point, we can outline the general problem that this chapter deals with, namely:

> *How to discover a common set of non-obvious causal structures that differentiate less and more efficient entities?*

We can also identify a population for which a solution to the problem could be of benefit, which is a:

> *A sample of convenience comprised of efficiency-driven entities, or groups of entities, that could be described by the same "Input → Output" production model and the same set of contextual attributes.*

Arguably, the process of benchmarking with the purpose of increasing the level of efficiency could be viewed through the lens of hypothetico-deductive logic (Godfrey-Smith, 2003) for, fundamentally, the process of benchmarking is causally deterministic in nature (e.g., impact C to get D), conclusion driven (e.g., change from *Conclusion: Inefficient* to *Conclusion: Efficient*), and based on the attempt to alter the state of minor premise to impact the conclusion. This perspective is illustrated in Table 22.2.

**TABLE 22.2**
**Example of Hypothetico-Deductive Logic (HDL) Perspective on Benchmarking**

| Steps of HDL | General Example | Example of "Training → Reorders" Causal Structure |
|---|---|---|
| Major Premise | Efficient entity N has a causal structure A with a state X | Efficient sales team N annually spends $700/salesperson on training and has 50 reorders per salesperson |
| Minor Premise | Inefficient entity M has a causal structure A with a state Y | Inefficient sales team M annually spends $200/salesperson on training and has 10 reorders per salesperson |
| Conclusion | Efficiency of Entity M could be impacted by the state Y of A | Efficiency of team M could be impacted by low spending on training |
| Insight | Entity M may benefit from changing the state of structure A from Y to X | Team M may benefit by spending more on training |

The importance of solving the general problem of this study is intuitive, for:

*Identifying a set of common causal structures that differentiate less and more efficient entities allows for obtaining insights regarding the ways of impacting the level of efficiency.*

As our reader can see, the purpose of this investigation is to find a way of discovering, basically, major and minor premises of the steps of hypothetico-deductive method. Let us offer the following illustrative example:

*Major premise*: A functionality (*Input–Output* process) N with the state A indicates relative efficiency.
*Minor premise*: Entity X has a functionality N with the state B.
*Conclusion*: Entity X is relatively inefficient.
*Problem*: Entity X must become relatively efficient – change *Conclusion* of HDL.
Benchmarking solution to the *Problem*: Entity X may change *Conclusion* by changing *Minor premise* – via matching *Minor premise* to *Major premise* and obtaining functionality N with the state A, not B.

Consequently, we put forward a set of propositions, expressed as follows:

*Proposition 1.1: The Process of benchmarking relies on the rules of hypothetico-deductive logic.*
Justification: The process of benchmarking relies on matching the actual state (*Minor premise*) of an entity with the desired state (*Major premise*) of an entity with the intent of changing the resultant behavior of the entity (*Conclusion*).
*Proposition 1.2: The problem of process inefficiency could be addressed by benchmarking.*
Justification: The state of an entity could be depicted by the process of conversion of its *Inputs* into *Outputs*.
*Proposition 1.3: Success in addressing the problem of process inefficiency via benchmarking is impacted by the discovery of a proper model of conversion of Inputs* into *Outputs.*
Justification: A model of conversion of *Inputs* into *Outputs* is not limited to pure *Inputs* and *Outputs*, but must also take into consideration relevant contextual factors.
*Proposition 1.4: A proper model of conversion of Inputs* into *Outputs could be expressed as a combination of two models – the "true" process model (dictated by the "true" production process, or given by the subject-matter expert), and the "contextual" process model (to be discovered in the case of context-dependency of the "true" process) that impacts, directly or indirectly, the "true" model.*
Justification: An enterprise is an open dynamic complex system that impacts, and is impacted by its context – competitive business environment (Amagoh, 2008; Samoilenko & Osei-Bryson, 2007a; Samoilenko, 2008a).

*Proposition 1.5: The overall success of the benchmarking effort is impacted by the discovery of a relevant contextual process model which could be expressed in the form of non-obvious causal structures.*

Justification: If a context impacts the process of conversion of *Inputs* into *Outputs*, then it is important to depict the context properly.

To formalize the process of discovery of causal structures we express our inquiry in a form of a methodology – this allows our readers to formally review and evaluate our approach.

The dynamic nature of the global business environment implies that as the time goes by and the competitive landscape changes the membership of the sample will also change and the causal structures will have to be discovered anew. Thus, we offer our solution to the stated above problem in the form of inductive methodology relying on non-parametric data analytic techniques and methods. Conceptually, we present the scope of the research problem that our methodology addresses in Figure 22.1.

Prior to describing our methodology, we would like to present to our reader a conceptual argument for why the proposed methodology should exist at all. A simple way of doing it is by presenting a set of propositions so that our readers could follow along the logic of our argument in a step-by-step fashion.

Proposition 2.1: Deductive approaches to process improvement (e.g., benchmarking) relies on the bounded and well-defined universe of discourse (UoD).

Proposition 2.2: Deductive approaches to process improvement are based on the rules of Hypothetico-Deductive Logic (HDL).

Proposition 2.3: Bounded and well-defined UoD is expressed as a set of Major and Minor premises of HDL, alteration of which results in changes to Conclusion of HDL.

Proposition 2.4: Changes in the dynamic business environment force re-definition of the UoD.

Proposition 2.5: In order to apply hypothetico-deductive approaches to process improvement in the context of changing business environment a new model of UoD must be created, re-bounded, and re-defined.

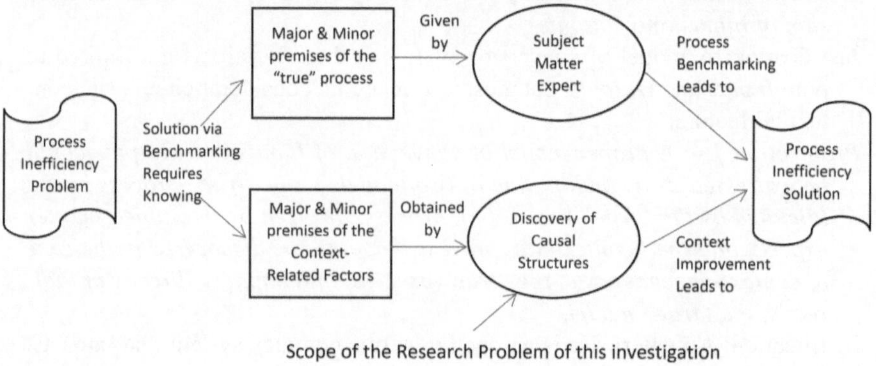


**FIGURE 22.1** Scope of the research problem.

Proposition 2.6: A new model of UoD reflecting a new business environment cannot be created using deductive (general → specific, current → past) methods, but can be created using inductive (specific → general, past → future) methods.

Proposition 2.7: Data mining and non-parametric methods allow for creating the new model of UoD via identifying changes that took place in the business environment.

Overall, this set of propositions could be visually expressed in the form of a Methodological Spiral depicted in Figure 22.2 which outlines the scope of our inquiry, and to set the limits of what our methodology can do. We summarize the applicability of our approach as follows:

Limitation 1: Our methodology is not applicable to static business environments, for such environment will not require re-defining UoD. Our methodology will be of greater use to firms that develop on-line services and provide outsourced customer support, that to the firms that build airports, oil pipelines, and chip manufacturing facilities.

Limitation 2: Our methodology is not applicable to the pure production process that is context –independent – not impacted by the business environment. For example, a process of manufacturing of Advil according to the given formula will not be aided by our methodology, while a process of a delivery of on-line orders will be.

Limitation 3: Our methodology is of limited importance to the process improvement approaches that are inductive in nature. Fundamentally, deductive approaches are based on copying of the best practices in changing contexts, and our methodology can offer valuable insights regarding changes that took place. Inductive approaches, such as business process re-engineering, are not based on copying of the best practices, but rather on creating unique best practices.

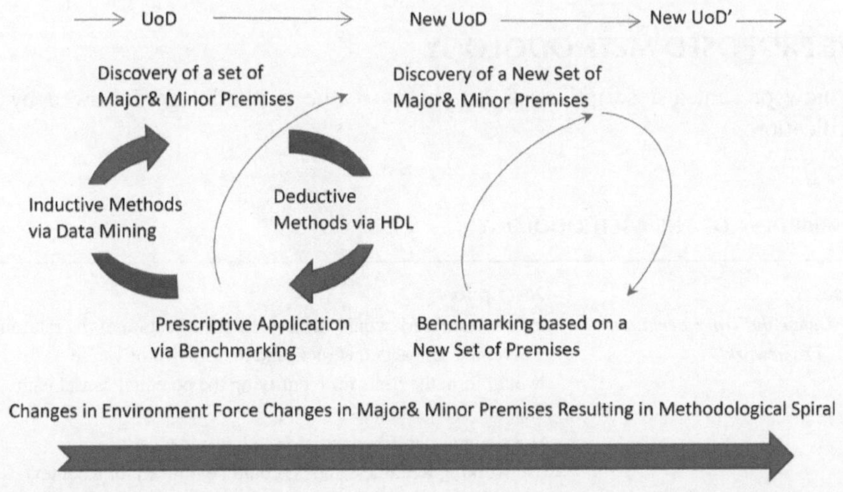


**FIGURE 22.2** Methodological spiral as a response to the changes in business environment.

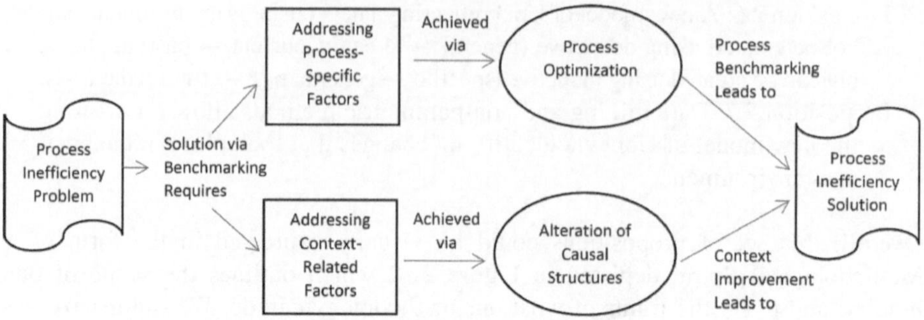


**FIGURE 22.3** Benchmarking as a generalized solution to the problem of process inefficiency.

Keeping this in mind, we can generalize the solution to the problem of process inefficiency via benchmarking as consisting of two elements. First, addressing process-specific factors, and, second, addressing context-related factors. The proposed solution is presented in Figure 22.3.

The problem of relative inefficiency could be tackled using a two-step approach. The first step deals with identifying relatively less-efficient and relatively more-efficient counterparts (e.g., *Leaders* vs. *Followers*). The second step deals with inquiring into the factors that could be associated with the relative inefficiency and then implementing changes to the factors. This approach works well if the scope of the undertaking is limited to the input–output conversion process. Sometimes, however, the roles are assigned based on external to the input–output process' factors, when, for example, less affluent economies try to benchmark their wealthier counterparts. In any case, the evaluation of the efficiency of the process of the conversion of inputs into outputs takes place.

## THE PROPOSED METHODOLOGY

We now present a description of the phases of the methodology followed by its justification.

### DESCRIPTION OF THE METHODOLOGY

| Phase | Description |
|---|---|
| 1 *Define the Transformation Framework* | Such a framework would involve the specification of the relevant Driver and Impact constructs and their indicator variables. It would form the basis for identifying the potential causal paths that are to be evaluated.<br>An existing established transformation framework (e.g., Networking Readiness Index) could be utilized, or a context-relevant model could be developed using logical informed arguments & insights from relevant existing Literature |

| Phase | | Description |
|---|---|---|
| 2 | *Partition the set of Decision Making Units into Meaningful Groups* | This may be done using a previously established rules for group membership (e.g., World Bank classification scheme), or the application of a data mining-based clustering process on a data set includes the *Input* (i.e., Drivers), *Mediator* and *Output* (i.e., Impacts) variables of the Transformation Framework |
| 3 | *Benchmark the DMUs* | For each group:<br>For each potential Causal Path, apply the DEA method to the data set in order to identify classify the DMUs in each DMU group in terms of the following: (a) Efficient vs. Non-Efficient; and (b) Growth in Productivity vs. Non-Growth in Productivity. |
| 4 | *Identify the Major Attributes that Differentiate the DMUs* | For each group:<br>• Apply Decision Tree Induction (DTI) in order to determine the major (i.e., top) attributes that determine the status of each DMU in terms of Relative Efficiency.<br>• Apply DTI in order to determine the top attributes that determine the status of each DMU in terms of Growth in Productivity. |
| 5 | *Discover the Causal Structures* | For each group:<br>• Apply Association Rules Mining (ARM) to discover the significant rules that describe the causal structures that are associated with: (a) Efficient vs. Non-Efficient; and (b) Growth in Productivity vs. Non-Growth in Productivity. |

## Justification & Benefits of the Methodology

The first phase of our methodology is associated with defining, or adapting, a context-specific transformation framework according to which inputs are converted into outputs. For all intents and purposes, the goal of the transformation framework is to create a set of established pathways by which the transformation of inputs into outputs takes place. We refer to the transformation framework as the *Universe of Discourse of Transformation (UoDoT)* – an established common set of means by which inputs could be transformed into outputs. A meaningful for our purposes *UoDoT* is *semi-constrained* in the sense that it exists between completely *constrained* and *unconstrained* states. Let us consider an illustrative example.

Let us consider Joe and Bob, two people who have some money to invest with the purpose of receiving a profit. So they do so and Joe does better than Bob. The basic high-level model for Joe and Bob is "Money → Investment → Profit" and the basic goal is the maximization of the outputs – profit. The first and the last parts of equation, *money* and *profit*, are perfectly clear – Joe and Bob have money, and they want to get more money. The problematic part is the second one – that of *investment*, because until Joe and Bob know what, exactly, it stands for, they cannot optimize the input–output process of "money → profit". And it is *investment* part in this example that we refer to as UoDoT, and it deals with the question of what is it, exactly, Joe and Bob could invest their money to get, as a result, more money. In this example, Bob does not do as well as Joe, and Bob asks for an advice – turns out all Joe does, just as Bob does, is to keep his money on his savings account, and Joe does better because his bank has better rates than Bob's bank. This is an example of *constrained UoDoT* – a single path (e.g., savings account) according to which money transformed into more money – revenue. Conversely, let us consider Joe not sharing his investment strategies with Bob–Joe could be investing in illegal drug trade, or in an underground amphetamine lab, or he could be really lucky playing state lottery, or Joe could be doing million other things. This is an example of *unconstrained UoDoT* – many different paths could exist, and no set of paths is known as the actual one. As our reader can see *constrained UoDoT* and *unconstrained UoDoT* do not allow for benchmarking to take place – in the former case there is nothing to benchmark, and in the latter case it is not known what to benchmark.

The situation changes, however, if we have a *semi-constrained UoDoT* where Joe invests his money in bonds, stocks, and hedge funds. In this case, the basic high-level model for Joe and Bob is "money → {bonds, stocks, and hedge funds} → profit", and Bob can benchmark Joe by learning intricacies of his investment strategies – when and how much to invest into what and under what conditions. It seems like the Joe and Bob scenario works fine for the illustrative purposes, but it leaves out one important point, namely, how does Bob know that Joe does better than him? Or, putting it in a general way: *how does an entity identifies its benchmarking target*? The second part of our methodology is designed to provide an answer to this question. It is during the second phase a data sample is partitioned into various sub-groups with the purpose of identifying a better performing entity or sub-group (benchmarking target), and a worse performing entity or sub-group, and various levels of entities or sub-groups in between. This could be done according to a given scheme (e.g., classification of IMF based on the income level of economies), or expert knowledge (e.g., general manager identifies top-, medium-, and low-performance retail stores), or by discovery via data mining methods. Needless to say, all the members of the sample must share the same *semi-constrained UoDoT*.

DEA is a method that is widely used for the purposes of calculating scores of the relative efficiency of entities that receive inputs and produce outputs. For example, we could compare three groups of basketball players of different levels (e.g., high school, college, and professional) in terms of their efficiency of conversion of minutes played and attempts taken into assists and points. Results of DEA would yield the most relatively efficient group, but because DEA model is a "black box" model we would not know what differentiates the groups, or why one group is more efficient than the other two.

The insights could be provided by Decision Tree Induction (DTI), which would yield an attribute, or a few attributes, that differentiate the groups. Thus, given a set of attributes describing the three groups of basketball players, we may find out that the main difference between the groups is in years of experience and hours of weekly practice. However, there are plenty of players with many years of experience who train long hours every week, but don't play so well. We would like to know what set of attributes is actually associated with the outputs – assists or points scored. Here is where ARM may help.

ARM allows for generating sets of association rules of the type "If (a,b,c) Then (d)". This is very valuable, for we can see the patterns of associations specific to each group. However, ARM tends to generate many rules, some trivial, some meaningless/non-actionable, and some useful. The problem with selecting the rules describing the different groups is that the rules may contain completely different attributes – this would result in comparing apples and oranges. For example, in the case of basketball players we may get "If (height > x) Then (minutes_played > y" for high school players, "If (experience > n) Then (assists > m)" and so on. So, the trick is to identify a set of rules that is based on a set of common criteria that differentiates the groups – this insight is provided by DTI. Consequently, the novelty of our approach is associated with its capability to identify the main differentiating factors responsible for heterogeneity of the context, and then to base the selection of the rules on those factors.

Previous investigations used a hybrid DEA/DTI methodology (Samoilenko & Osei-Bryson, 2008), and the use of ARM with DEA was recently reported by Samoilenko (2016); however, this investigation represents the first case of using the three methods (i.e., DEA, DTI, ARM) *in synergy*. Simply put, if DEA allows us to identify the efficient performers, and DTI helps us to discover the relevant dimensions that differentiate efficient and inefficient performers, then ARM allows us to benchmark efficient performers via a set of "If → Then" rules that rely on the dimensions discovered by DTI analysis. To our knowledge, no other combination of data analytic and data mining methods could offer so much in so few steps.

Banker & Natarajan (2008) had also presented a 2-stage DEA+OLS method which assumes that contextual variables have been identified but they provided no guidance as to how such variables could be identified. Appropriate application of Banker & Natarajan's and similar methods in a real-world context would thus require that before such methods are applied that appropriate activities to factor in context be conducted. Phases 1 & 2 of our proposed framework describe such a set of activities.

Worthington & Dollery (2002) had suggested that "environmental (or contextual) factors may encompass both physical environmental circumstances, as well as constraints arising from organizational and managerial policies". Later Barbosa, Lima & Brusca (2016) noted that: "the scores of efficiency are calculated from a set of inputs and outputs; however, the explanatory variables of these models do not represent inputs or outputs, but the context where each DMU is". In this chapter, we consider the concept of context in a manner similar to that of Worthington & Dollery (2002) and Barbosa, Lima & Brusca (2016); we consider the context to involve non-discretionary explanatory variables that are not used as inputs in the calculation of the efficiency scores; contextual variables could be categorical, ordinal or interval. Thus, for example, given a heterogeneous set of DMUs, the group to which a given DMU belongs is an aspect of the context in which it operates as it converts its inputs to outputs. Since as noted by Worthington & Dollery (2002) and other researchers that ignoring context "may lead to disingenuous efficiency measures" we present a multi-phase framework that that aims to provide guidance for the identification of the aspects of context before applying DEA.

Banker & Natarajan (2008) demonstrated that under a specific set of conditions (e.g., "monotone increasing and concave production function separable from a parametric function of the contextual variables") a 2-stage DEA+OLS model yields "consistent estimators of the impact of the contextual variables". It is not clear if such conditions typically apply in real-world contexts, but when they do apply then their 2-stage DEA+OLS model could be applied in what corresponds to Phases 3 & 4 of our framework to identify the variables that impact productivity. Interestingly, the use of DTI to replace OLS in the stage model does not require the pre-requisite conditions assumed by Banker & Natarajan (2008).

## ILLUSTRATIVE EXAMPLE – APPLICATION TO SUB-SAHARAN ECONOMIES

Our illustrative example involves the application of the methodology to a data set that covers a sub-set of the countries of Sub-Saharan Africa (SSA). We obtained the data from a publicly available source – World Economic Forum' Global Information Technology Report (GITR, 2015). In 2012 the representation of the NRI has partially changed, so we decided to concentrate on the new version of NRI and use the data for 2012–2015 period. However, for some SSA economies data was missing for some years (e.g., Angola, Seychelles, Liberia, Gabon, Sierra Leone, and Guinea) and represents 27 SSA economies.

### Phase 1: Define the Transformation Framework

In our illustrative example, we will use the NRI framework (Dutta, Geiger & Lanvin, 2015), the adapted version of which is depicted below in Figure 22.4. The framework relies on four subindexes and their ten sub-categories (or pillars) to obtain the value of the NRI, which reflects the capacity of economies to benefit from ICT. An increase in the value of NRI for a given economy is indicative of the increase in the impact of ICT on innovation and productivity (Dutta & Jain, 2003).

Interestingly, the original framework does not explicitly connect Environment, Readiness, and Usage subindexes (referred to as Drivers within the framework) with Impact subindex (referred to as Impact), despite relying on a principle that "…the environment, readiness, and use—interact, co-evolve, and reinforce each other to create greater impact" (Di Battista, Dutta, Geiger & Lanvin, 2015, p.4).

We scope our inquiry by only considering relationships between Drivers (environment, readiness, and use) and Impact (socio-economic impact of Drivers) – as indicated by arrows in Figure 22.4. All possible interactions within Drivers (e.g., between environment, readiness, and use) we consider to be beyond the scope of our investigation. In our inquiry, we use the framework depicted in Figure 22.4 to

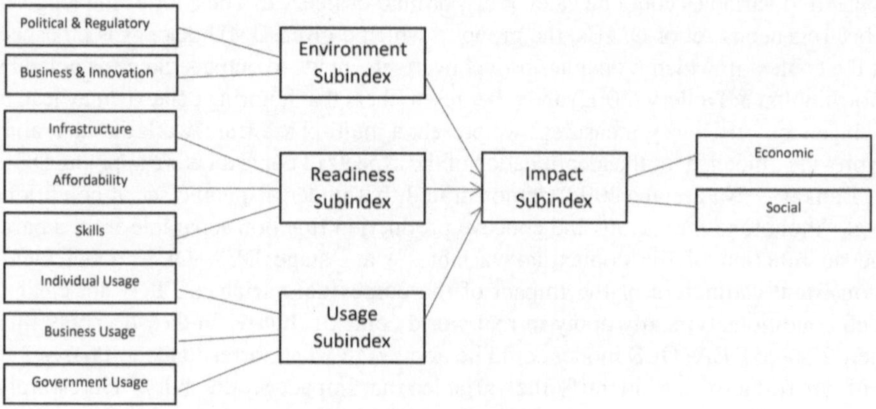


**FIGURE 22.4** The framework of networked readiness (adapted format).

investigate, via DEA, the efficiency of the process by which Environment, Readiness, and Usage subindexes impact two sub-categories of Impact subindex – Economic and Social impacts.

## Phase 2: Partition the Set of Decision Making Units into Meaningful Groups

Our set of DMUs are countries of SSA. In some of our previous work (e.g., Self-Reference) that involved the transition economies (TEs) of Eastern Europe, we used cluster analysis to partition the set of TEs into two meaningful groups. While that option is available, so are other options include World Bank classification schemes. So, in this study, we will use the classification of the World Bank as of July 2014 to partition the set of SSAs into three groups: *Low Income*, *Low-Middle Income* & *Upper-Middle Income* (see Table 22.3).

We obtained the data from a reputable source – the World Economic Forum's Global Information Technology Report 2015 (GITR, 2015). In some cases, the representation of SSA economies was inconsistent, for example, we could not include Angola, Seychelles, Liberia, Gabon, Sierra Leone, and Guinea in our sample because the data for some of the years was missing. While there is an advantage to increasing the sample size of a study, there is a price to pay via dealing with missing variables, imputation of values, and additional data preprocessing. After considering the pluses and minuses of "sample size vs. data actually available", we have assembled a smaller data set that contained no missing data and no outliers, but was as reliable as one could get from a given source. Overall, we were able to compile the data set representing 27 SSA economies (the classification of the International Monetary Fund as of October 2014).

## Phase 3: Data Envelopment Analysis

During the first phase, we rely on DEA to evaluate relative efficiency of three "Drivers → Impact" paths. We will use *variable return to scale* (VRS) DEA model to conduct the analysis, for it is reasonable to argue that SSA economies have not yet reached the point of developing a level of ICT infrastructure allowing accruing the benefits yielded by capitalizing on economies of scale.

**TABLE 22.3**
**Sample of Sub-Saharan African (SSA) Economies, by Income Level**

| Income Level | Sub-Saharan Africa (SSA) Economies |
|---|---|
| *Low Income* | Burkina Faso, Burundi, Chad, Ethiopia, Gambia, Kenya, Madagascar, Malawi, Mali, Mozambique, Rwanda, Tanzania, Uganda, Zimbabwe |
| *Low-Middle Income* | Cameroon, Cape Verde, Côte d'Ivoire, Ghana, Lesotho, Nigeria, Senegal, Swaziland, Zambia |
| *Upper-Middle Income* | Botswana, Mauritius, Namibia, South Africa |

**TABLE 22.4**
**DEA Models of the Illustrative Example**

| DEA Model | Inputs of DEA Model | Outputs of DEA Model |
|---|---|---|
| *Drivers* → Overall_*Impact* (DOI) | Environment Subindex<br>Readiness Subindex<br>Usage Subindex | Impact Subindex |
| *Drivers* → Economic_*Impact* (DEI) | Environment Subindex<br>Readiness Subindex<br>Usage Subindex | Economic Impact Sub-category |
| *Drivers* → Social_*Impact* (DSI) | Environment Subindex<br>Readiness Subindex<br>Usage Subindex | Social Impact Sub-category |

Given a four-years time period, we will run DEA 12 times. Consequently, for each economy in the sample, we are going to have four scores of relative efficiency for each of the three models (see Table 22.4). At this point, we need to provide a justification for the inputs and outputs included in our models (Cook, Tone & Zhu 2014).

In regard to outputs, the reasoning is intuitive – first, we would like to assess the efficiency of the overall impact, and then, each type of the impact separately. This is because an economy could be efficient in obtaining one type of an impact (e.g., economic) and not efficient in regard to another impact (e.g., social).

With regard to the choice of the inputs of DEA model, our approach is methodological. While we are free to use eight sub-categories of Drivers as inputs of a DEA model, the general rule of thumb is that for a reasonable level of discrimination number of economies (or *Decision Making Units* in DEA terms) must be at least twice the product of inputs and outputs (Dyson, Allen, Camanho, Podinovski, Sarrico & Shale, 2001). In our case we have a sufficient number of economies in our set, but if we use a DEA model with eight inputs and two outputs then we would need to have at least $2 \times 8 \times 2 = 32$ economies in the sample. Furthermore, and more importantly, the greater the number of factors included in the DEA model, the lower the level of discrimination of the model (Dyson et al., 2001).

Our results, summarized in Table 22.5, demonstrated that seven economies of the full sample are relatively efficient with regard to the impact of Drivers on social, economic, and overall impact of ICT.

Additionally, we identified relatively efficient economies per each of the income-level group; in some cases (e.g., Burundi, Chad, Kenya, Mali, Rwanda, and Senegal), the relatively efficient within its group' economies are also efficient overall. In other cases (e.g., Swaziland, Lesotho, Botswana, Mauritius, Namibia, and South African Republic), the relatively inefficient, overall, economies end up being efficient within their respective group.

Also, we used DEA to calculate the values of Malmquist index (*MI*), which allows us to assess the changes in the scores of relative efficiency that took over period of time. Under the assumption of constant returns to scale the change indicates changes

**TABLE 22.5**
**Results of DEA**

| Income Level | Economy | Overall Efficiency of the Impact of ICT | Overall Changes in Productivity | Growth via EC? | Growth via TC? |
|---|---|---|---|---|---|
| LI | BFA | Efficient | Growth | Yes | Yes |
| | BDI | Efficient | No growth | Yes | No |
| | TCD | Efficient | No growth | No | No |
| | ETH | Efficient | No growth | No | No |
| | GMB | Inefficient | No growth | Yes | No |
| | KEN | Efficient | No growth | No | No |
| | MDG | Inefficient | Growth | Yes | No |
| | MWI | Inefficient | No growth | No | No |
| | MLI | Efficient | Growth | No | Yes |
| | MOZ | Inefficient | No growth | Yes | No |
| | RWA | Efficient | No growth | Yes | No |
| | TZA | Inefficient | Growth | Yes | No |
| | UGA | Inefficient | No growth | Yes | No |
| | ZWE | Inefficient | Growth | Yes | No |
| LM | CMR | Inefficient | Growth | Yes | No |
| | CPV | Inefficient | No growth | No | No |
| | GHA | Inefficient | No growth | No | No |
| | NGA | Inefficient | No growth | No | No |
| | SEN | Efficient | No growth | No | Yes |
| | SWZ | Inefficient | Growth | Yes | No |
| | ZMB | Inefficient | Growth | Yes | No |
| | CIV | Inefficient | Growth | Yes | No |
| | LSO | Inefficient | Growth | Yes | No |
| UM | BWA | Inefficient | No growth | No | No |
| | MUS | Inefficient | Growth | Yes | No |
| | NAM | Inefficient | No growth | Yes | No |
| | ZAF | Inefficient | No growth | No | No |

in productivity. Consequently, we are able to assess whether the economies become more productive or not. Because *MI* is comprised of two components – change in efficiency (*EC*) and change in technology (*TC*), we are also able to assess whether the changes in productivity are associated with a particular component. Overall, only 11 economies (40% of the sample) exhibited growth in productivity.

An analysis of the changes in *EC* and *TC* offers an interesting insight: 16 economies (60% of the sample) exhibited positive changes in efficiency, but only 3 economies (11% of the sample) demonstrated positive changes in technology. Finally, it is worth noting that only one economy, Burkina Faso, exhibited a balanced growth in productivity, when the growth was driven by both components of *MI*. Overall, the picture suggests that SSA, as a group, would benefit from a better technology – this suggests that investments in ITC infrastructure should be prioritized.

Finally, it is worth noting that 8 economies (30% of the sample) have not only exhibited a decline in productivity but have exhibited a decline in terms of both components of MI.

## Phase 4: Decision Tree Induction (DTI)

To proceed with this phase, we need to create a new variable "Target" to differentiate various groups of economies. We are interested in three types of groupings: first, we would like to differentiate the SSA groups by their level of economic development, and then we would like to differentiate the groups in terms of their efficiency of the socioeconomic impact of ICT. The last analysis would involve differentiating SSA economies by growth in productivity – *Growth vs. No Growth.* Thus, we would conduct DTI three times, which would require Target to have three domains of values.

In the first case, grouping by income, the domain of values of *Target* would be {1, 2, 3}, for, respectively, *Low Income* (LI), *Low Middle Income* (LM), and *Upper Middle Income* (UM) groups of economies. In the second case, grouping by efficiency, The Target would assume the values of {0, 1}, for, respectively, *relatively inefficient*, and *relatively efficient* SSA. The same domain of values, namely, {0, 1}, could be applied to the grouping by growth in productivity, where "0" would indicate "No Growth" and "1" would indicate "Growth".

The results of DT analysis summarized in Table 22.6 show that such pillars of NRI as *Individual Usage*, *Business Usage*, and *Skills Readiness* do play important role in differentiating three groups of economies. It is not surprising that there appears to be a clear-cut difference between Low Income and Upper-Middle Income economies, and much less of a difference between Low-Middle Income economies and the other two.

We could also identify *Individual Usage* and *Economic Impact* as pillars that play role in differentiating relatively efficient SSA economies from inefficient ones. It appears that Infrastructure Readiness, Affordability Readiness, and Individual Usage are factors playing role in differentiating those economies that became more productive from those that didn't.

**TABLE 22.6**
**Results of Decision Tree Analysis**

| Grouping by | Group | Differentiating/Split Variable |
|---|---|---|
| Economic development | Low Income vs. Low Middle Income vs. Upper Middle Income | Individual usage<br>Business usage<br>Skills readiness |
| Relative efficiency | Relatively Efficient vs. Relatively Inefficient | Individual usage<br>Economic impact |
| Change in productivity | Growth vs. No Growth | Infrastructure readiness<br>Affordability readiness<br>Individual usage<br>Social impact |

## PHASE 5: ASSOCIATION RULE MINING

The purpose of this phase is to find possible patterns, associations, or causal structures that may exist in our data. One of the main advantages of association rule mining (ARM) is that it is suitable for undirected data mining; thus, we'll aim to discover naturally occurring associations between the factors (sub-indexes of *Drivers* and *Impact*) – components of NRI. ARM could be classified as either being explanatory or exploratory in nature. In the case of our investigation, we employ exploratory ARM, for we do not have any theoretical support for why certain relationships between the sub-indexes of NRI should exist. A very common approach to generating associations between the variables, or *itemsets*, via ARM is by using the *a priori* algorithm (Agrawal & Ramakrishnan, 1994) – we will rely on this approach in the current investigation. Transformation of the data is required for this step – we follow the method of Samoilenko & Osei-Bryson (2017) to do so.

The application of ARM result in the generation of multiple association rules (ARs) that contain context-specific combinations of factors (see Table 22.7).

The results of the data analysis offered in the previous sections offer evidence that we were successful in developing and testing a novel methodology that provides for

**TABLE 22.7**
**Impact-Specific Rules for Low Income and Low-Middle Income Economies**

| Condition | Generated Rules | | | Sup. | Conf. | Lift |
|---|---|---|---|---|---|---|
| Low Income | low IND_USE, low BUS_USE | => | low ECON_IMP | 21% | 0.70 | 1.9 |
| | low SKILL_READ, low BUS_USE | => | low ECON_IMP | 21% | 0.60 | 1.5 |
| Low-Middle | midhigh BUS_USE | => | midlow SOCIO_IMP | 20% | 0.50 | 2.5 |
| | midhigh SKILL_READ | => | midhigh SOCIO_IMP | 25% | 0.50 | 1.5 |
| Upper-Middle | high SKIL_READ, high AFFORD_READ, high GOV_USE | => | high SOCIO_IMP | 32% | 1.00 | 2.3 |
| | high BUS&INNOV_ENV, high INFR_READ, high BUS_USE | => | high ECON_IMP | 31% | 0.85 | 2.3 |
| | high BUS&INNOV_ENV, high IND_USE, high BUS_USE | => | high ECON_IMP | 31% | 0.85 | 2.2 |
| Low Income, Inefficient | low INFR_READ, low BUS_USE | => | low SKILL_READ | 36% | 1.00 | 1.9 |
| | low IND_USE, low BUS_USE | => | low SKILL_READ | 32% | 1.00 | 1.9 |
| Low-Middle, Efficient | low BUS_USE, low GOV_USE | => | low ECON_IMP | 25% | 1.00 | 4.0 |
| | low BUS_USE, low ECON_IMP | => | low SOCIO_IMP | 25% | 1.00 | 3.0 |

investigating complex context-specific relationships between the factors reflecting the state and the impact of *ICT Capabilities*. The discussion of the results is presented along the points that we considered noteworthy.

First,

> Despite the presence of complex relationships between the Drivers and Impacts of ICT there are common themes associated with the levels of the scores of factors comprising NRI – *Business Usage, Individual Usage*, and Skills Readiness appear to have a direct relationship with the levels of the scores of socio-economic Impact of ICT.

We point out that while the variety of association rules has been generated for a different set of criteria, a common line could also be glanced – some sub-categories of NRI' subindexes (e.g., related to Skills, Business, Individual usage) appear more frequently than other sub-categories.

Second,

> These results suggest that *Business Usage* and I*ndividual Usag*e are among the factors that appear to differentiate economies in terms of their level of economic development, as well as in terms of their relative efficiency of the impact of ICT on the socioeconomic bottom line.

These results suggest that wealthier and more efficient economies tend to have higher scores of *Business Usage* and *Individual Usage*. The presence of a simple association between the level of income of an economy, its efficiency, and ICT usage seems to be apparent.

Third,

> The *Infrastructure* and *Affordability* of ICT seems to have an impact on growth in productivity of SSA economies.

This finding is important because it was corroborated, independently, by two different methods of analysis – DEA and DT.

## CONCLUSION

In this chapter, we have presented a new multi-method methodology for benchmarking that uncovers relevant context-specific causal structures. We then provided an illustrative application of this methodology to an IS/ICT & Productivity research problem in the "developing" countries context that resulted in the uncovering of causal structures that describe relationships between *Drivers* and *Impact* of ICT. These results show that the relationships between the factors representing *Drivers* and *Impact* are indeed complex – in the sense of Complex Systems Theory where structure of the systems changes (Samoilenko, 2008a) as a response to the pressures of the environment (e.g., only some of the pillars of NRI differentiate the clusters) and the relationships between the components are non-linear (e.g., the association rules are not stable, but level-dependent). However, the underlying complexity of these relationships can be made more transparent to researchers and practitioners by the application of the proposed methodology. This increase in transparency comes, unsurprisingly, from elucidation of the role of the context within which a complex dynamic system, be it an economy, or an industry, or a firm, operates.

While the other commonly used DEA-based methodologies may allow for uncovering the sources of heterogeneity between the various groups in the sample (e.g., Samoilenko & Osei-Bryson, 2007b), or for differentiating on purely efficiency-based reasons for discrepancies in the levels of performance (e.g., Samoilenko & Osei-Bryson, 2013), our methodology also allows for discovering (unlike the approach of Banker & Natarajan (2008) that takes the "received knowledge" approach) actionable causal structures that are context-based and setting-specific, and in this regard the proposed in this chapter approach is truly unique. In summary, there are several reasons on which claims for the novelty and value of this methodology can be based including:

- It subsumes the 2-stage DEA framework of Banker & Natarajan (2008).
- It does not require the pre-requisite conditions (e.g., "monotone increasing and concave production function separable from a parametric function of the contextual variables") assumed by Banker & Natarajan (2008).
- It has the capability to identify the main differentiating factors responsible for heterogeneity of the context.
- It allows for obtaining actionable information, in the form of non-obvious common causal structures, for improving the performance of the less efficient entities vis-à-vis their more efficient counterparts
- It allows for allows for expanding the universe of discourse within which the process improvement initiatives are usually considered, thus allowing to consider the impact of external to the process factors on internal to the process mechanisms.
- It involves a new & creative integration of multiple data mining methods (CA, DTI, ARM) with DEA.

## ACKNOWLEDGMENT

Material in this chapter previously appeared in: A data analytic benchmarking methodology for discovering common causal structures that describe context-diverse heterogeneous groups. *Expert Systems with Applications*, *117*, 330–344.

## REFERENCES

Adler, N., Liebert, V., and Yazhemsky, E. (2013). Benchmarking Airports from a Managerial Perspective. Omega, 41(2), 442–458, https://doi.org/10.1016/j.omega.2012.02.004.

Agrawal, R. and Ramakrishnan S. (1994). Fast Algorithms for Mining Association Rules in Large Databases. In Proceedings of the 20th International Conference on Very Large Data Bases (VLDB '94), Jorge B. Bocca, Matthias Jarke, and Carlo Zaniolo (Eds.). Morgan Kaufmann Publishers Inc., San Francisco, CA, 487–499.

Al-Mashari, M., Irani, Z., and Zairi, M. (2001). Business Process Reengineering: A Survey of International Experience. Business Process Management Journal, 7(5), 437–55.

Amagoh, F. (2008). Perspectives on Organizational Change: Systems and Complexity Theories. The Innovation Journal: The Public Sector Innovation Journal, 13(3), 1–14.

Ayabakan, S., Bardhan, I. R., and Zheng, Z. (2017). A Data Envelopment Analysis Approach to Estimate IT-Enabled Production Capability. MIS Quarterly, 41(1).

Banker, R.D. and Natarajan R. (2008). Evaluating Contextual Variables Affecting Productivity Using Data Envelopment Analysis. Operations Research, 56(1), 48–58.

Barbosa, A., Lima, S. C. D., and Brusca, I. (2016). Governance and Efficiency in the Brazilian Water Utilities: A Dynamic Analysis in the Process of Universal Access. Utilities Policy, 43(PA), 82–96.

Cao, G., Clarke, S., and Lehaney, B. (2001). A Critique of BPR from a Holistic Perspective. Business Process Management Journal, 7 (4), 332–339.

Cook, W., Tone, K., and Zhu, J. (2014). Data Envelopment Analysis: Prior to Choosing a Model. Omega, 44, 1–4, https://doi.org/10.1016/j.omega.2013.09.004.

Di Battista, A., Dutta, S., Geiger, T., and Lanvin, B. (2015). The Networked Readiness Index 2015: Taking the Pulse of the ICT Revolution. Global Information Technology Report 2015, 3–28.

Dutta, S. and Jain, A. (2003). The Networked Readiness of Nations. In The Global Information Technology Report 2002–2003, Dutta, S., A. Lanvin, B., and Paua, F. (eds.). Oxford University Press, New York, NY, Oxford.

Dutta, S., Geiger, T., and Lanvin, B. (Eds.) (2015). Global Information Technology Report 2015. ICTs for Inclusive Growth. World Economic Forum and INSEAD, Geneva. Retrieved April 20, 2015 from http://www3.weforum.org/docs/WEF_Global_IT_Report_2015.pdf

Dyson, R.G., Allen R., Camanho A.S., Podinovski, V.V., Sarrico C.S., and Shale, E.A. (2001). Pitfalls and Protocols in DEA. European Journal of Operational Research, 132, pp. 245–259.

Fui-Hoon Nah, F., Lee-Shang Lau, J., and Kuang, J. (2001). Critical Factors for Successful Implementation of Enterprise Systems. Business Process Management Journal, 7 (3), 285–296.

GITR. (2015). World Economic Forum' Global Information Technology Report. Available on-line at: http://reports.weforum.org/global-information-technology-report-2015/network-readiness-index/.

Godfrey-Smith, P. (2003). Theory and Reality: An Introduction to the Philosophy of Science. University of Chicago Press.

Gouveia, M.C., Dias, L.C., Antunes, C.H., Boucinha, J., and Inácio, C.F. (2015). Benchmarking of Maintenance and Outage Repair in an Electricity Distribution Company using the Value-based DEA Method. Omega, 53, 104–114, https://doi.org/10.1016/j.omega.2014.12.003.

Hitt, L. M. and Brynjolfsson, E. (1996). Productivity, Business Profitability, and Consumer Surplus: Three Different Measures of Information Technology Value. MIS Quarterly, 121–142.

Ko, M. and Osei-Bryson, K. M. (2004). Using Regression Splines to Assess the Impact of Information Technology Investments on Productivity in the Health Care Industry. Information Systems Journal, 14(1), 43–63.

Korotin, V., Popov, V., Tolokonsky, A., Islamov, R., and Ulchenkov, A. (2017). A Multi-criteria Approach to Selecting an Optimal Portfolio of Refinery Upgrade Projects under Margin and Tax Regime Uncertainty. Omega, 72, 50–58, https://doi.org/10.1016/j.omega.2016.11.003.

LaPlante, A.E. and Paradi, J.C. (2015). Evaluation of Bank Branch Growth Potential using Data Envelopment Analysis. Omega, 52, 33–41, https://doi.org/10.1016/j.omega.2014.10.009.

Merriam-Webster. (2017). Benchmark. Retrieved October 26, 2017, from https://www.merriam-webster.com/dictionary/benchmark

Moon, Y.B. (2008). Enterprise Resource Planning (ERP): A Review of the Literature. International Journal of Management and Enterprise Development, 4 (3), 235–64.

Samoilenko, S. and Osei-Bryson, K.M. (2007a). Chaos Theory as a Meta-Theoretical Perspective for IS Strategy: Discussion of the Insights and Implications. In Proceedings of the Southern Association for Information Systems Conference, Jacksonville, FL, May 22–23, 2007.

Samoilenko, S. and Osei-Bryson, K.M. (2007b). Increasing the Discriminatory Power of DEA in the Presence of the Sample Heterogeneity with Cluster Analysis and Decision Trees. Expert Systems with Applications, 34(2), 1568–1581.

Samoilenko, S. (2008a). Information Systems Fitness and Risk in IS Development: Insights and Implications from Chaos and Complex Systems Theories. Information Systems Frontiers, 10(3), 281–292.

Samoilenko, S. (2016). Disparity of Social and Economic Impact of ICT Capabilities in Sub-Saharan Economies: Empirical Investigation of Differentiating Factors. In Proceedings of the SIG GlobDev Pre-ECIS Workshop ICT in Global Development, Istanbul, Turkey, June 12, 2016.

Samoilenko, S. and Osei-Bryson, K.M. (2008). An Exploration of the Effects of the Interaction between ICT and Labor Force on Economic Growth in Transitional Economies. International Journal of Production Economics, 115, 471–481.

Samoilenko, S. and Osei-Bryson, K.-M. (2013). Using Data Envelopment Analysis (DEA) for Monitoring Efficiency-Based Performance of Productivity-Driven Organizations: Design and Implementation of a Decision Support System. Omega, 41(1), 131–142.

Samoilenko, S. and Osei-Bryson, K.-M. (2017). A Methodology for Identifying Sources of Disparities in the Socio-Economic Impacts of ICT Capabilities in Sub-Saharan Economies. In Proceedings of the 2017 International Conference on Information Resources Management (conf-IRM 2017). http://aisel.aisnet.org/confirm2017/2

Techinsights.com. (2018). Apple IPhone 8 Plus Teardown. TechInsights, techinsights.com/about-techinsights/overview/blog/apple-iphone-8-teardown/.

Taylor, F. W. (1914). The Principles of Scientific Management. Harper.

Worthington, A. C. and Dollery, B. E. (2002). Incorporating Contextual Information in Public Sector Efficiency Analyses: A Comparative Study of NSW Local Government. Applied Economics, 34(4), 453–464.

# 23 An Empirical Investigation of ICT Capabilities and the Cost of Business Start-up Procedures in Sub-Saharan African Economies

Investments in information and communication technologies (ICTs) are common to all economies of the world, and they are allocated to pursue common goals of positively impacting context-specific socio-economic targets. One of the underlying principles of the framework is that "NRI should provide clear policy guidance" (p. xi) to its users, which is not an easy undertaking to implement because within the framework "…the complex relationships between ICTs and socio-economic performance are not fully understood and their causality not fully established" (Di Battista, Dutta, Geiger, & Lanvin, 2015, p. 4). Simply put, *Drivers* of the framework are not directly associated with *Impact*. This is not surprising, for *ICT Capabilities*, reflected by *Drivers*, are simply opportunities that must be, in one form or another, taken advantage of in order to produce any socio-economic *Impact*.

Perhaps, an illustration may help: an individual may have a set of capabilities qualifying her to work (e.g., level of education, level of experience, and level of social maturity), but this set of capabilities does not directly result in any level of socio-economic impact (e.g., salary, title, level of seniority, and social status). Instead, in this case the proper chain of links representing "Capabilities → Impact" will be represented by the sequence "Capabilities → Employment → Impact". Thus, capabilities, or *drivers*, must fuel some sort of an *Engine* that produces an *Impact*. Similarly, within the framework of NRI ICT capabilities reflected by *Drivers* must enable some sort of a socio-economic *Engine* that actually produces *Impact*.

It is because of this absence of the direct association between *Drivers* and *Impact*, that the framework offers a great platform for exploring various economic means that power social and economic well-being of any economy. The overall framework for such exploration, consistent with the framework of NRI, is represented in Figure 23.1. It is worth noting that this framework is sufficient enough to investigate any type of *Engine* of the impact of *ICT Capabilities*, and it is minimal enough in allowing doing so in the most efficient manner. Furthermore, this framework is consistent with a grand economic theory – the framework of Neoclassical Growth Accounting (Solow, 1957).

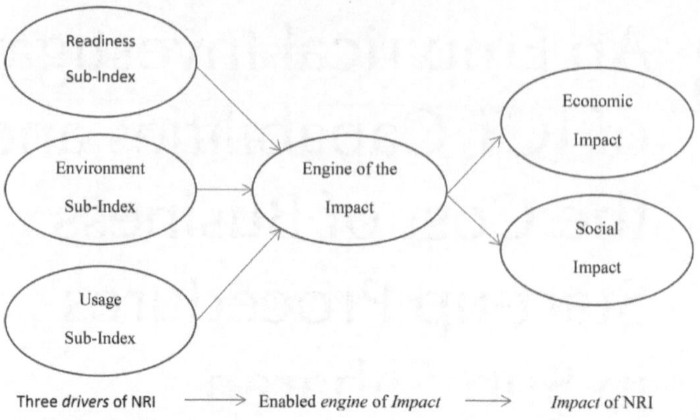


**FIGURE 23.1** Framework for exploration of socio-economic impacts of ICT Capabilities.

According to the Neoclassical model, long-term economic growth is a product of capital investment, labor, and technological progress. To reconcile the framework of Figure 23.1 with the Neoclassical model, we only need to accept the fact that *ICT Capabilities* (e.g., *drivers*) of NRI are not given, but a result of appropriate investments in ICT. Furthermore, if we allow for a certain portion of socio-economic impact to be re-invested, then we will arrive at a framework of sustainable macroeconomic impact of investments in ICT similar to that of Samoilenko and Osei-Bryson (2017). The resultant framework is depicted in Figure 23.2.

One of the recognized engines of socio-economic impact is *Small and Medium Enterprises* (SME), which is not only a source of over 50% of formally documented jobs in the world but also is a provider of effective solutions to critical development issues (TWB, 2016). The presence of SMEs is essential to the poorest countries in the world, those of *Sub-Saharan Africa* (SSA), where they serve as an important driver of economic growth that accounts for a majority of all businesses (IFC, 2017).

Furthermore, African SMEs are also important to the global economy because their presences and success create "…a growing middle class with disposable income, in tandem with market opportunities for new investors" (de Sousa dos Santos, 2015)

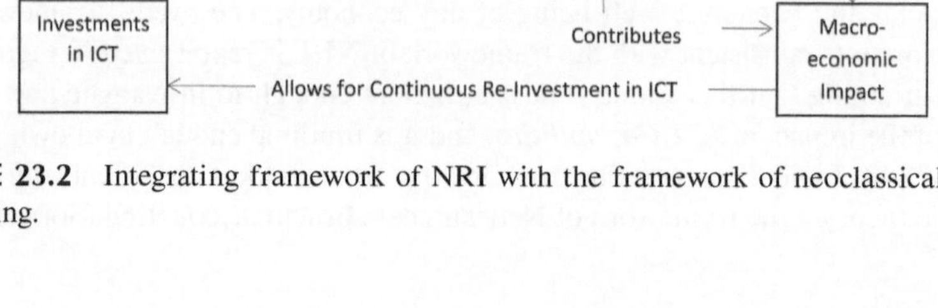


**FIGURE 23.2** Integrating framework of NRI with the framework of neoclassical growth accounting.

and it is expected that over the next two decades Sub-Saharan economies "...will become the main source of new entrants into the global labor force" (IMF, 2015).

It was noted, however, that while the SMEs of SSA are dominant players in their economies, they typically run unregistered businesses (Fierro, 2015). This has two important implications. First, informal enterprises do not participate in the formal economy, and, second, they have a very limited access to finance (Fjose, Grünfeld, & Green, 2010). Both implications are important, but it is an access to finance that was identified as a critical factor for the survival and growth of SSA's SMEs (Fierro, 2015).

There are various factors that may prevent informal enterprises from becoming registered SMEs, but we call the attention of our readers to an obvious one – *Cost of Business Start-Up Procedures (CBSP)* in a given economy. The justification for choosing this factor is an intuitive one, for *CBSP must be incurred in order to formally register an SME.* The importance of this factor is also straightforward, for *formally registered SMEs have an easier access to finance required for sustaining and developing the business.*

In this chapter we investigate two broad questions within the context of SSA:

1. *Whether ICT Capabilities impact the Cost of Business Start-Up Procedures (CBSP)?*
2. *Whether the cost of business start-up procedures is associated with a socio-economic impact?*

The conceptual underpinnings of these questions are simple. Let us consider the role of ICT and of the *ICT Capabilities* in any economy – their primary role is to create new information channels and/or to increase the efficiency and effectiveness of the existing ones. By fulfilling this role, ICT *should* reduce transaction costs and, as a result, decrease the *CBSP*. But does it do so effectively? Does it do so efficiently? We look closely at these questions in this chapter.

There are at least two types of socio-economic impacts that *CBSP* can deliver. First one is associated with the *consequences* of the legalization of the business and all possible positive results of it. This type of an impact is a complex one, for it must account for a wide variety of benefits brought to all the beneficiaries of the legalization within the business (e.g., owners, employees, and other stakeholders) and within the context (e.g., social community, vendors, and customers) where the business operates. We consider the first type of impact to be outside of the scope of this chapter. The second type of impact is a much simpler one and is associated with a pure economic cost of legalization of the business. This cost is incurred by an individual, or individuals, and must be allocated out of the disposable income or a loan. It is precisely this type of impact that we investigate in this study.

Let us now consider *CBSP*, in its pure and simple form of incurring associated fees, as an engine of socio-economic impact. First of all, it is only expected that a lower cost of legalization of SME produces direct economic impact, for if less money is allocated toward the start-up procedures, then more money will be available for business development or toward out-of-pocket spending. Similarly, we expect that the low *CBSP* will produce a positive social impact, for a registered business owner, *ceteris paribus*, enjoys a higher level of social recognition than an owner of unregistered business. Additionally, lower *CBSP* would leave more money to be allocated toward education and health care. But does, indeed, *CBSP* produce a socio-economic impact? We aim to answer this question in our inquiry.

Essentially, we approach the subject of this inquiry from the perspective of a complex adaptive system, where the target system (e.g., a given SSA economy or group of economies) is comprised three components (e.g., *ICT Capabilities*, *CBSP*, and socio-economic outcomes) bound by non-linear relationships. And it is, fundamentally, the context-dependent strength of the relationships is of primary concern to this study. At this point we can formulate the overall question that we aim to answer, as follows:

*Whether ICT Capabilities can result in a positive socio-economic outcome by impacting the CBSP in the context of SSA economies?*

We conduct our investigation within the context of 26 economies of SSA, using the data set for the period of 2012–2016. The analysis of the data is supported by a five-phase methodology utilizing cluster analysis (CA), data envelopment analysis (DEA), decision tree induction (DTI), association rule mining (ARM), and ordinary least squares (OLS).

## THE RESEARCH FRAMEWORK AND RESEARCH QUESTIONS

In our investigation we rely on the NRI framework (Kirkman, Osorio, & Sachs, 2002), the adapted version of which is depicted below in Figure 23.3. The framework relies on four sub-indexes and their ten sub-categories (or *pillars*) to obtain the value of the NRI, which reflects the capacity of economies to benefit from ICT.

This research framework allows for answering a set of important questions. We offer the formulation of the questions (*Q#:*), along with the rationale for each question (*R#:*), below.

*Q1: Do SSA economies in our sample represent a homogenous group in regard to NRI and CBSP?*

R1: We know that SSA in the sample belongs to three different income groups; however, this differentiation is a received one (e.g., based on classification of World Bank), and is not based on our study' factors (e.g., NRI and CBSP).

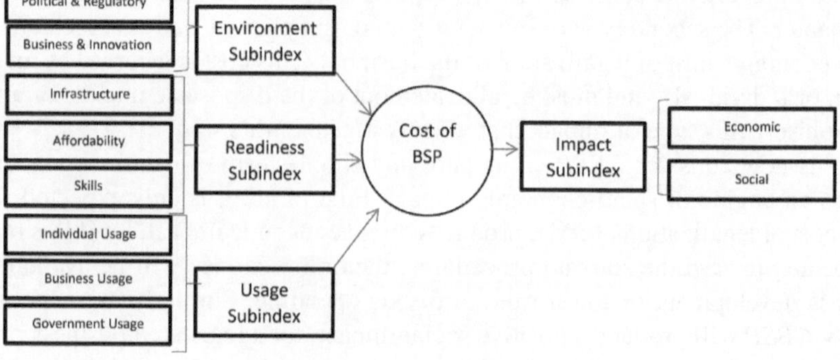


**FIGURE 23.3** The research framework of the study.

*Q2: If the sample of SSA economies comprises multiple sub-groups, then what differentiates the heterogeneous sub-groups?*

R2: Knowing what differentiates heterogeneous sub-groups allows for additional information to be used in decision making; specifically, we can identify the sources of heterogeneity in the sample.

*Q3: In the case of the sample heterogeneity, how do sub-groups of SSA economies differ in terms of the relative efficiency of the conversion of ICT Capabilities into CBSP?*

R3: Knowledge of the relative efficiency of the conversion of *ICT Capabilities* into *CBSP* for each of the groups allows us to identify the best performers among the groups of SSA economies. This information is not only relevant to the research in the area of ICT4D, but it is also invaluable for the practical purposes of benchmarking and sharing the best practices in the context of SSA.

*Q4: Is there a set of actionable "If → Then" rules that could be generated to differentiate the best performers from the other groups in the sample?*

R4: By discovering a set of actionable "If → Then" rules that include common factors laggards will be able to benchmark the best performers in the sample. Furthermore, discovering naturally occurring causal structures allows for obtaining deeper insights regarding the mechanisms of socio-economic impact of ICT.

*Q5: Is there a relationship between the value of CBSP and socio-economic impact?*

R5: By answering this question we will know whether the groups of SSA economies differ in regard to the impact of *CBSP*.

## PROPOSED METHODOLOGY

The aim of this part of the chapter is to describe the purpose and the expected outcome of each of the five phases of the methodology, while leaving the technical details of each method out. The methodology of this investigation is an extension of the methodology of Samoilenko and Osei-Bryson (2016).

### Phase 1: Cluster Analysis (CA)

The purpose of CA is to identify naturally occurring sub-sets, or groupings, that may exist in the data set. While in our case SSA economies are categorized by the level of income, that categorization is based on the external criteria and is not *CBSP*- or NRI-specific. By performing CA, we can determine if SSA economies are alike in terms of NRI- and *CBSP*-specific criteria or if the sample is heterogeneous and consists of multiple sub-groups. The outcome of Phase 1 is N-cluster solution. It is worth noting that partitioning of the sample into sub-groups is logical in nature (Samoilenko and Osei-Bryson, 2008), where no physical division of the sample takes place.

### Phase 2: Decision Tree Induction

If the results of Phase 1 yield a multi-cluster solution, then we create a variable *target*, with the domain of values being the number of identified clusters (e.g., 3-cluster solution will yield *target* {1,2,3}). By performing DTI, we will be able to identify, based on the top-level splits, the variables that differentiate the clusters the most. The outcome of Phase 2 is a set of variables differentiating N-clusters comprising the sample.

### Phase 3: Data Envelopment Analysis

During the third phase, we rely on DEA to evaluate the scores of relative efficiency of the "Drivers → CBSP" paths for each of N clusters in the sample. The purpose of using DEA in this investigation is to evaluate the level of efficiency with which *ICT Capabilities* transformed into *CBSP.* The outcome of Phase 3 is a set of averaged scores for each of N clusters, which would allow us to identify the benchmark group in the sample of SSA economies. We also conduct DEA to assess the relative efficiency of each of the clusters in terms of conversion of *drivers* into *impact.* We will use two models- the first one to assess the overall impact and the second to assess social and economic impact.

### Phase 4: Ordinary Least Squares Regression

The purpose of Phase 4 is to evaluate the presence of the causal relationships between *CBSP* and socio-economic outcomes. For each of N clusters of the sample, we will run OLS three times with the purpose of regressing *CBSP* against social, economic, and socio-economic impacts of NRI. The outcome of Phase 4 is a set ($3 \times N$ clusters) of p-values allowing for the testing of the significance of the relationships.

### Phase 5: Association Rule Mining

The purpose of Phase 5 is to find possible patterns, associations, or causal structures that may exist in our data. One of the main advantages of ARM is that it is suitable for undirected data mining; thus, we'll aim to discover naturally occurring associations between the pillars of NRI and *CBSP* that may describe, in "If → Then" fashion, sub-groups of our sample. The outcome of Phase 5 is a set of actionable "If → Then" rules incorporating the variables discovered in Phase 2.

## DATA

We obtained the data from two publicly available and highly reputable sources. NRI' data was downloaded from World Economic Forum' Networked Readiness Index' page (WEF_NRI, 2016) and the values of *CBSP (% of GNI per capita)* were obtained from the Data Bank of the World Bank (The World Bank, 2016).

In 2012 the number and representation of the *pillars* of three sub-indexes, the *drivers*, of NRI has changed, and it was also the year when the *impact* sub-index was

**TABLE 23.1**
**Sample of Sub-Saharan Economies, by Income Level**

| Income Level | SSA Economies |
|---|---|
| Low | Burundi, Chad, Ethiopia, Gambia, Kenya, Madagascar, Malawi, Mali, Mozambique, Rwanda, Tanzania, Uganda, Zimbabwe |
| Low- Middle | Cameroon, Cape Verde, Côte d'Ivoire, Ghana, Lesotho, Nigeria, Senegal, Swaziland, Zambia |
| Upper- Middle | Botswana, Mauritius, Namibia, South Africa |

introduced. Given the changes that took place between 2011 and 2012, we decided to concentrate on the new version of NRI and collect the data for the period between 2012 and 2016. The inclusion of SSA economies in the WEF' *Global Information Technology Report* (the source of the annual NRI data) is not always consistent. Resultantly, we could not include some of the SSA economies (e.g., Angola, Burkina Faso, Seychelles, Liberia, Gabon, Sierra Leone, and Guinea) in our sample simply because the data for some of the years were missing.

Overall, we were able to compile the data set representing 26 economies of Sub-Saharan Africa (the classification of the International Monetary Fund as of October 2014). The sample consists of 13 low-income economies, nine low-middle economies, and four upper-middle economies (the classification of the World Bank as of July 2014). Membership of each group of the sample is provided in Table 23.1.

## RESULTS OF THE DATA ANALYSIS

### Phase 1: CA

While performing CA we have followed a rule of thumb for the relative size of any single cluster being greater than 10% of the overall size of the sample (Samoilenko and Osei-Bryson, 2008). By performing hybrid hierarchical clustering (there seems to be a consensus that this approach combines the advantages of other methods (Laan & Pollard, 2002; Chipman & Tibshirani, 2006) without having any unique disadvantages), we arrived at a 3-cluster solution. The results are summarized in Table 23.2.

Based on the original solution, 19 SSA economies are "permanent residents" of a given cluster, while 7 SSA economies are "migrants". This indicates that 77% of the clusters are stable in regard to their membership. Furthermore, if we consider four out of five years to be the indicator of stability, then over 92% of the clusters' membership is stable.

### Phase 2: DTI

By performing DTI, we were able to identify sub-indexes and pillars that differentiate the three clusters of SSA economies in our sample. During this phase we constructed various decision trees, based on different variables (e.g., sub-indexes,

**TABLE 23.2**
**Results of Clustering: Membership, Income Composition, and Size of Each Cluster**

| Cluster # | Composition | Cluster Membership |
|---|---|---|
| **Cluster 1:** **Size:32, A_CBSP = 29.05** | LI 21.88%<br>LM 43.75%<br>UM 34.38% | Mauritius, South Africa, Cape Verde, Ghana, Zimbabwe (2013–2015), Botswana (2012), Uganda (2012–2015), Kenya (2014–2016) |
| **Cluster 2:** **Size: 62, A_CBSP = 57.90** | LI 69.35%<br>LM 30.65%<br>UM 0.00% | Burundi, Chad, Ethiopia, Madagascar, Malawi, Mali, Mozambique, Tanzania, Swaziland, Côte d'Ivoire (2012–2014), Nigeria (2015–2016), Uganda (2016), Lesotho (2012–2015), Zimbabwe (2012, 2016) |
| **Cluster 3:** **Size: 36, A_CBSP = 49.43** | LI 27.78%<br>LM 47.22%<br>UM 25.00% | Cameroon, Senegal, Zambia, Gambia, Rwanda, Namibia, Kenya (2012–2013), Côte d'Ivoire (2015–2016), Nigeria (2012–2013), Botswana (2013–2016) |

pillars, NRI, and CBSP, level of income). We were able to determine, for example, that the three clusters cannot be clearly differentiated based on the income level, or based on the value of NRI, or based on *impact* sub-index of NRI. However, we obtained evidence that the clusters do differ based on the values of *drivers* (sub-indexes), and the values of the associated pillars. We present summary of the most relevant results of DTI in Table 23.3.

**TABLE 23.3**
**Results of DTI: English Rules, Split Variables, and Split Criteria**

| English Rules | Split Variables |
|---|---|
| **Pillars-based DT** | B04: Readiness sub-index: Affordability |
| B04 > 3.880 | B05: Readiness sub-index: Skills |
| \| B05 > 2.815: One {three=0, **one=32 (100%)**, two=0} | A01: Environment Sub-index: Political and regulatory environment |
| \| B05 ≤ 2.815: two {three=1, one=0, two=10} | C07: Usage sub-index: Business usage |
| B04 ≤ 3.880 | |
| \| C07 > 3.220 | |
| \| \| A01 > 3.005: three {**three=35 (97%)**, one=0, two=5} | |
| \| \| A01 ≤ 3.005: two {three=0, one=0, two=4} | |
| \| C07 ≤ 3.220: two {three=0, one=0, **two=43 (69.35%** } | |
| **Sub-index-based DT** | |
| B > 3.725: One {three=0, **one=29 (90.63%),** two=0} | A- Environment sub-index |
| B ≤ 3.725 | B- Readiness sub-index |
| \| C > 2.950 | C- Usage sub-index |
| \| \| A > 3.530: three {**three=34 (94.44%),** one=2, two=1} | |
| \| \| A ≤ 3.530: two {three=0, one=1, two=5} | |
| \| C ≤ 2.950: two {one=0, three=2, **two=56 (90.32%)**} | |

## Phase 3: DEA

During the third phase, we rely on DEA to evaluate the relative efficiency of "Drivers → CBSP" path for each of the three clusters. We will use *variable return to scale* (VRS) DEA model to conduct the analysis, for it is reasonable to argue that SSA economies have not yet reached the point of developing a level of ICT infrastructure allowing for accruing the benefits yielded by capitalizing on economies of scale. We also conduct DEA to evaluate the relative efficiency of "Drivers → Overall Impact" and "Drivers → (Social Impact, Economic Impact)" paths (Table 23.4).

One point worth noting is the way of handling the variable *Cost of Business Startup Procedures.* DEA allows us to use three common models: input-, output-, and base-oriented. Input orientation deals with minimizing the level of inputs to achieve a given level of output. Output orientation is concerned with maximizing the level of outputs based on the given level of inputs. Base orientation refers to having control over inputs and outputs at the same time. In our case, we use an output-oriented model, for we are interested in the output-side of the DEA model. However, we are interested in minimizing CBSP, not in maximizing it. To deal with this situation we used a simple conversion scheme – first, we rounded up the greatest value of CBSP in the sample and labeled it *CBSP Max*. Then, for each data point, we subtracted the actual value of CBSP from *CBSP Max* and used that value in DEA. Results of the analysis presented in Table 23.5.

It is interesting to note that while there is practically no difference between the clusters in regard to two *Drivers → Impact* models, there is a major difference between cluster 2 and the other two clusters in terms of *Drivers → CBSP* and two *CBSP → Impact* models.

## Phase 4: OLS

In this phase of our methodology, we run a regression analysis for each of the three clusters of SSA economies in our sample. For each cluster, we create three OLS

**TABLE 23.4**
**DEA Model of the Study**

| DEA Model | Inputs of DEA Model | Outputs of DEA Model |
|---|---|---|
| *Drivers → CBSP* | Environment Sub-index<br>Readiness Sub-index<br>Usage Sub-index | Cost of Business Startup Procedures |
| *CBSP → Overall Impact* | CBSP | Impact Sub-index |
| *CBSP → (Social Impact, Economic Impact)* | CBSP | Social Impact Pillar<br>Economic Impact Pillar |
| *Drivers → Overall Impact* | Environment Sub-index<br>Readiness Sub-index<br>Usage Sub-index | Impact Sub-index |
| *Drivers → (Social Impact, Economic Impact)* | Environment Sub-index<br>Readiness Sub-index<br>Usage Sub-index | Social Impact Pillar<br>Economic Impact Pillar |

**TABLE 23.5**
**Results of DEA**

| DEA Model | Model Orientation | Relative Efficiency Scores, Averaged | | |
|---|---|---|---|---|
| | | Cluster 1 | Cluster 2 | Cluster 3 |
| *Drivers → CBSP* | *Output-Oriented* | 1.24 | 4.49 | 1.58 |
| *CBSP → Overall Impact* | *Input-Oriented* | 0.27 | 0.006 | 0.05 |
| *CBSP → (Social Impact, Economic Impact)* | *Input-Oriented* | 0.31 | 0.006 | 0.19 |
| *Drivers → Overall Impact* | *Input-Oriented* | 0.93 | 0.94 | 0.97 |
| *Drivers → (Social Impact, Economic Impact)* | *Input-Oriented* | 0.96 | 0.96 | 0.97 |

model, regressing the original value of CBPS against (1) socio-economic impact (sub-index), (2) social impact (pillar of *impact* sub-index), and (3) economic impact (pillar of *impact* sub-index). The results of OLS are summarized in Table 23.6.

## Phase 5: ARM

Overall, we ran ARM analysis four times. The first time we analyzed the complete sample (all three clusters) for the presence of general "If → Then" rules that would identify conditions for low-level *CBSP*. The results are provided in Table 23.7. We would like to remind our readers that the results of CA and DTI identified the following four variables impacting the heterogeneity of the sample:

- *Affordability Readiness* pillar (*Readiness* sub-index)
- *Skills Readiness* pillar (*Readiness* sub-index)
- *Political and regulatory Environment* pillar (*Environment* sub-index)
- *Business Usage* pillar (*Usage* sub-index).

**TABLE 23.6**
**Results of OLS**

| Cluster | Depend. Var. | Adj. *R* | *F*-value | *P*-value (95%) | Significant? |
|---|---|---|---|---|---|
| 1 | Socio-economic Impact | 0.351 | 17.772 | **0.000** | Yes |
| | Social Impact | 0.209 | 9.172 | **0.005** | Yes |
| | Economic Impact | 0.339 | 16.864 | **0.000** | Yes |
| 2 | Socio-economic Impact | –0.014 | 0.150 | 0.700 | No |
| | Social Impact | –0.006 | 0.617 | 0.435 | No |
| | Economic Impact | –0.016 | 0.025 | 0.875 | No |
| 3 | Socio-economic Impact | 0.120 | 5.790 | **0.022** | Yes |
| | Social Impact | 0.045 | 2.652 | 0.113 | No |
| | Economic Impact | 0.139 | 6.658 | **0.014** | Yes |

**TABLE 23.7**
**Results of ARM, Complete Data Set**

| Left Side (If) | & | → | Right Side (Then) | Sup. | Conf. | Lift |
|---|---|---|---|---|---|---|
| **HIGH Political & Regulatory Environment**<br>HIGH Infrastructure Readiness | & | → | LOW CBSP | 0.14 | 0.86 | 3.38 |
| **HIGH Political & Regulatory Environment**<br>HIGH Individual Usage | & | → | LOW CBSP | 0.14 | 0.86 | 3.38 |
| **HIGH Political & Regulatory Environment**<br>**HIGH Skills Readiness** | & | → | LOW CBSP | 0.11 | 0.93 | 3.68 |
| **HIGH Political & Regulatory Environment**<br>HIGH Infrastructure Readiness<br>HIGH Individual Usage | & | → | LOW CBSP | 0.13 | 1.00 | 3.94 |
| HIGH Infrastructure Readiness<br>**HIGH Skills Readiness**<br>HIGH Individual Usage | & | → | LOW CBSP | 0.12 | 0.88 | 3.48 |

Consequently, we can identify the association rules that contain the variables (in bold) listed above.

Results provided in Table 23.7 suggest that the level of CBSP is associated with the levels of two pillars – *Political & Regulatory Environment* and *Skills Readiness*. Because in this case ARM analysis was conducted using the complete data set, we have a basis to generalize that such association is may be valid for all SSA economies in our sample.

Next, we ran ARM analysis three times – once for each of the sub-groups in the sample. By selecting the rules that contain variables that differentiate the sub-groups, we were able to identify level-dependent associations that offer important actionable insights. We offer the summary of the results in Table 23.8.

**TABLE 23.8**
**Results of ARM, Summarized Comparisons of Three Clusters**

| Left Side (If) | & | → | Right Side (Then) | Sup. | Conf. | Lift |
|---|---|---|---|---|---|---|
| **HIGH Political & Regulatory Environment** | | → | LOW CBSP | 0.34 | 0.85 | 1.69 |
| **MIDLOW Political & Regulatory Environment** | | → | MIDHIGH CBSP | 0.19 | 0.88 | 2.10 |
| **HIGH Political & Regulatory Environment**<br>HIGH Individual Usage | & | → | LOW CBSP | 0.31 | 0.83 | 1.67 |
| **MIDLOW Political & Regulatory Environment**<br>MIDHIGH Individual Usage | & | → | MIDHIGH CBSP | 0.19 | 0.88 | 2.10 |
| **HIGH Skills Readiness**<br>**HIGH Business Usage** | & | → | HIGH EconImpact | 0.31 | 0.91 | 2.08 |

*(Continued)*

**TABLE 23.8**
**(Continued)**

| Left Side (If) | & | → | Right Side (Then) | Sup. | Conf. | Lift |
|---|---|---|---|---|---|---|
| **LOW Business Usage** LOW Government Usage | & | → | LOW EconImpact | 0.29 | 1.00 | 1.94 |
| **HIGH Business Usage** | | → | HIGH EconImpact | 0.34 | 0.92 | 2.10 |
| **LOW Political & Regulatory Environment** **LOW Business Usage** | & | → | LOW EconImpact | 0.23 | 1.00 | 1.94 |
| **HIGH Business Usage** HIGH Government Usage | & | → | HIGH SocImpact | 0.25 | 0.90 | 1.80 |
| **LOW Business Usage** LOW Government Usage | & | → | LOW SocImpact | 0.27 | 0.94 | 1.95 |
| **HIGH Political & Regulatory Environment** HIGH Government Usage | & | → | HIGH SocImpact | 0.25 | 1.00 | 2.00 |
| **LOW Political & Regulatory Environment** **LOW Business Usage** | & | → | LOW SocImpact | 0.23 | 1.00 | 2.07 |
| **HIGH Business Usage** | | → | HIGH SocImpact | 0.34 | 0.92 | 2.10 |
| **LOW Business Usage** LOW Government Usage | & | → | LOW SocImpact | 0.29 | 1.00 | 1.94 |

## INTERPRETATION OF THE RESULTS OF THE DATA ANALYSIS

In this section of the chapter, we offer to our readers a discussion of the results of our investigation. We proceed in the order of the steps of our methodology, where, for each step, we offer our interpretation of the results and briefly comment on the significance of the findings.

### Cluster Analysis

Despite sharing a common designation of SSA economies, the countries in our sample turned out to be quite different, and the difference was not along the lines of the received criterion (e.g., level of income) supplied by the World Bank. The three groups identified by the analysis turned out to be fairly stable in terms of their membership over a period of time. Interestingly, the membership of the sub-groups of the sample presented a mix in regard to the received categorization of the level of income (e.g., low income, low middle income, and upper middle income). There are three valuable insights that the results of CA provide to an investigator:

1. *The presence of the commonly shared designation of SSA economies should not be mistaken for the indication of heterogeneity of SSA economies*
2. *The sources of heterogeneity of SSA economies are specific to the area of inquiry*
3. *Any received categorization (such as via income level) must be rigorously tested if the area of inquiry is different than that of received categorization.*

## Decision Tree Induction

The obtained information regarding the heterogeneity of our sample of SSA economies led us to inquiring into the sources of heterogeneity – this was accomplished by performing DTI analysis. Based on the framework of the study, there are three possible sources of heterogeneity-based on the value of NRI, based on the values of sub-indexes of NRI, and based on the pillars of sub-indexes of NRI. Additionally, we decided to explore if the heterogeneity of the clusters could be partially explained by the level of income. The obtained evidence suggested that neither NRI nor the level of income is among the variables differentiating the sub-groups. However, sub-indexes and pillars of sub-indexes are turned out to be effective at explaining the differences between the sub-groups. We decided to concentrate on pillars of the sub-indexes of NRI, for such choice offered a lower level of granularity for our inquiry and, as a result, a greater discriminatory power in explaining the differences between the sub-groups. The decision is easy to justify – we identified that only *drivers*' sub-indexes and their respective *pillars* adequately explain the heterogeneity of the sample. Consequently, the choice was either to use three or eight variables – we chose the latter option.

Overall, out of 10 *pillars* of NRI, there are four that explain the differences between the sub-groups of SSA economies – it was unexpected to find out that the values of 60% of the *pillars* of NRI play no role in differentiating SSA economies in our sample. Furthermore, it was surprising to find out that the countries in our sample do not discernably differ in terms of social and economic impacts of ICT capabilities, and that the heterogeneity of the sample could be adequately explained in terms of only four variables – *Affordability Readiness*, *Skills Readiness*, *Political & Regulatory Environment*, and *Business Usage* pillars of NRI. These findings offered some important insights, as follows:

1. The economies of SSA do differ in a fairly narrow way – a sufficient explanation for the presence of heterogeneity in the sample can be offered on the basis of only four pillars (40% of the available variables) of the framework of NRI
2. The sub-groups of SSA economies do not differ significantly on the basis of socio-economic impacts of ICT Capabilities (the output side of "Input → Output" equation)
3. The sub-groups of SSA economies do differ on the basis of ICT Capabilities (the input side of "Input → Output" equation)
4. Based on general "Input → Output" model, SSA economies do not differ in terms of efficiency and effectiveness of production of outputs (socio-economic outcomes), but they do differ in terms of efficiency of utilization of inputs (ICT capabilities)
5. Overall, the insights suggest that the groups of SSA economies differ in terms of effectiveness and efficiency of transforming investments in ICT into a narrow sub-set (four pillars of NRI) of ICT capabilities (drivers of NRI).

## DEA

We arrived to the third step of our methodology knowing that the sub-groups comprising the sample of SSA economies do differ in terms of *ICT Capabilities* (e.g., *drivers* of NRI), and that they do not differ in terms of socio-economic outputs (e.g., *impact* of NRI). We also determined that the three sub-groups do differ in regard to the values of *CBSP* – the *Engine* of transformation of *Drivers* into *Impact.*

DEA evaluates the relative efficiencies of decision making units (DMU) – SSA economies – regarding conversion of inputs into outputs. By using five different DEA models as a test bed, we were able to obtain some valuable insights regarding the appropriate context models.

Perhaps, it is worthwhile to comment on what in our view constitutes an *appropriate* DEA model. Let us consider the fact that scores of relative efficiency of the DMUs calculated via DEA are derived from a ratio of inputs to outputs specified by the DEA model. Consequently, given the differences in initial conditions – the levels of values of inputs and outputs, it is only expected that the differences should be reflected by the scores of DEA. Meaning, a DEA model should be able to translate the differences in initial conditions (e.g., values of inputs and outputs) into the differences in the scores of relative efficiency that are based on initial conditions.

However, the results of DEA indicated that the three sub-group that do differ in regard to *ICT Capabilities* do not discernably differ in terms of the scores of relative efficiency of conversion of *ICT Capabilities* (e.g., *Drivers*) into socio-economic outcomes (e.g., *impact*). Resultantly, we consider "Drivers → Impact" model to be not appropriate for the context of our inquiry.

At the same time, we considered two other DEA models to be appropriate – the models are "Drivers → Engine" (e.g., *ICT Capabilities → CBSP*) and "Engine → Impact" (e.g., *CBSP → Socio-Economic Outcomes*). The first model, *ICT Capabilities → CBSP*, accurately reflected the differences in levels of inputs and outputs and the impact of the differences on the overall scores of relative efficiency. Insights provided by the results of the analysis suggest that SSA economies belonging to Cluster 2 should significantly lower the level of output – values of *CBSP*. Similarly, the second model, *CBSP → Socio-Economic Outcomes*, suggests that the members of Cluster 2 should or, lower the levels of *CBSP*, or increase the values of socio-economic outcomes. Given the fact that the values of *CBSP* are driven by policies, and not economic necessities, the wise policy decision advice to SSA economies – members of Cluster 2 is to significantly lower the costs associated with the business start-up procedures.

Overall, we summarize the importance of the findings provided by DEA as follows:

1. *If using DEA for the purposed of investigating a relative efficiency of the conversion of ICT Capabilities into Socio-Economic Outcomes two DEA models of "ICT Capabilities → Engine of transformation" and "Engine of Transformation → Socio-Economic Outcomes" are preferred to a single model of "ICT Capabilities → Socio-Economic Outcomes"*
2. *Overall, the relative efficiency of conversion of ICT Capabilities into Socio-Economic Outcomes is impacted by the levels of CBSP that serves as an engine of conversion of capabilities into outcomes.*

## Ordinary Least Squares (OLS)

After we performed CA, we calculated the average value of the *CBSP* for each of the three sub-groups of our sample. It turned out that Cluster 1 has the lowest; Cluster 2 has the highest, while Cluster 3 being in between but closer to Cluster 2 in regard to the average value of *CBSP*. The purpose of performing OLS analysis was to find out whether the different levels of values of *CBSP* impact socio-economic outcomes of *ICT Capabilities* differently. Simply put, we wanted to find out if lower levels of *CBSP* are better for SSA economies than the higher levels.

It worth pointing out that *any* value of *CBSP* that is greater than zero has a positive economic impact, for when start-up applicants pay the fee that money goes towards the economic bottom line. So, the question that we wanted to investigate is whether it is more beneficial to have an immediate and assured economic impact (e.g., via charging high fee), or if it is better to have a "delayed gratification" of charging less at the beginning and getting a greater socio-economic impact later when the money that could have been charged make their way to socio-economic outcomes.

As we expected, the results of OLS were intuitive – lower values of *CBSP* were positively impacting social, economic, and the socio-economic impacts of *ICT Capabilities*, while high values of *CBSP* have no discernable impact. The obtained insight is important, especially for policy-making purposes, for

1. *Higher levels of CBSP do not produce a discernable socio-economic impact, while the socio-economic impact of lower levels of CBSP is significant*
2. *One of the impacts of ICT Capabilities should be directed toward lowering of CBSP.*

## ARM

By the time we performed ARM analysis, we knew that our sample of SSA economics is comprised three sub-groups, and we also knew the factors differentiating the sub-groups. As a result, we were able to utilize ARM to identify two important sets of rules. The first set comprises general rules that were generated on the basis of the complete sample of SSA economies. This set allows us to identify the general patterns that seem to hold within our context. The second set contains cluster-specific rules that are based on the dimensions (discovered via DTI) differentiating the sub-groups. This set offers an opportunity to not only examine cluster-specific associations but also to compare the associations between the clusters.

Having such two sets of rules is important for three reasons. First, it allows for having two different perspectives – the global of the complete set and the local of each cluster. Second, it allows for merging of the two perspectives, where we can relate cluster-level rules to complete set-level rules. Third, it gives us an opportunity for meaningful local inter-cluster rule analysis while relating the analysis to the global set of rules.

Overall, we interpret the obtained via ARM analysis insights as follows:

1. *There is evidence that in the context of SSA economies high levels of Political & Regulatory Environment, Infrastructure Readiness, Skill Readiness, and Individual usage are associated with low levels of CBSP.*
2. *Consequently, if an SSA economy aims at having a low level of CBSP, then it is reasonable to suggest that investments in ICT should be directed at developing high-level capabilities in the areas of Political & Regulatory Environment, Infrastructure Readiness, Skill Readiness, and Individual Usage.*
3. *The results of ARM suggest that there is an inversely proportional relationship between the levels of Political & Regulatory Environment and CBSP.*
4. *There is evidence that the level of the economic impact of ICT Capabilities is directly proportional to the levels of Political & Regulatory Environment, Skill Readiness, and Business Usage.*
5. *There is also evidence that the level of the social impact of ICT Capabilities is directly proportional to the levels of Political & Regulatory Environment, Government Usage, and Business Usage.*

## DISCUSSION OF THE RESULTS OF THE STUDY

Prior to conducting inquiry, we identified five research questions that served as a set of stepping stones leading to answering two main questions of the study. At this point, we are well equipped to provide the answers to each of the research questions, as well as to offer some "lesson learned" type of insights.

*Q1: Do SSA economies in our sample represent a homogenous group in regard to NRI and CBSP?*

*A1: SSE economies do represent a homogeneous group in regard to the values of NRI, but they do differ in regard to the values of CBSP*

The obtained insight is intuitive – one should not assume the presence of homogeneity in the sample that is subjected to multi-dimensional analysis.

*Q2: If the sample of SSA economies is comprised heterogeneous sub-groups, then what differentiates the sub-groups?*

*A2: Sub-groups of SSA economies differ in regard to the values of four pillars of NRI: Affordability Readiness, Skills Readiness, Political & Regulatory Environment, and Business Usage*

The obtained in this regard insight is related to *dimensionality of the heterogeneity* of the sample, where despite the presence of multiple dimensions potentially relevant to the analysis, it is usually a small sub-set of variables that will be identified (or selected) as the pertinent one to explain heterogeneity of the sample.

*Q3:* *In the case of the sample heterogeneity, how do sub-groups of SSA economies differ in terms of the relative efficiency of the conversion of ICT Capabilities into CBSP?*

*A3:* *The sub-groups of SSA economies do differ in terms of the relative efficiency of the conversion of ICT Capabilities into CBSP, where the group with the high levels of ICT Capabilities and the low levels of CBSP is relatively more efficient than the group with the low level of ICT Capabilities and the high levels of CBSP*

It seems to be reasonable to suggest that if a DEA model contains variables that are also identified as sources of heterogeneity of the sample, then the heterogeneity of the sample will manifest in the results of DEA, thus yielding a sub-set of relatively efficient DMUs and a sub-set of relatively inefficient ones.

*Q4:* *Is there a set of actionable "If → Then" rules that could be generated to differentiate the best performers from the other groups in the sample?*

*A4:* *There is a set of "If → Then" rules that differentiate the best performers from the rest of the group- in general, the following summary rule seems to hold true: "(HIGH Political & Regulatory Environment, HIGH Skills Readiness, HIGH Business Usage) → (LOW CBSP, HIGH Economic Impact, HIGH Social Impact)"*

The insight obtained from answering this question is not surprising – given the fact that any DEA model is expressed in the form "inputs/outputs", it is of benefit to describe the difference between more and less relatively efficient groups of DMUs in the form of "(sub-set of *Inputs*) → (sub-set of *Outputs*)".

*Q5:* *Is there a relationship between the value of CBSP and socio-economic impact?*

*A5:* *Yes, there is a relationship between the value of CBSP and socio-economic impact and evidence suggests that SSA economies with low levels of CBSP tend to be associated with the high socio-economic impact, while SSA economies with the high levels of CBSP show no discernable socio-economic impact.*

The insight offered by this finding is a fairly important one, for it indicates that the mere presence of the *engine* of an impact (e.g., *CBSP*) does not guarantee the actual *impact*. Instead, the effectiveness of the engine of an impact is context-dependent and cannot be taken for granted.

We would like to remind our reader that the subject of this investigation, to put it simply, was an inquiry into what can be called "two impacts". First, we wanted to find out whether *ICT Capabilities* impact the *CBSP*. Second, we wanted to investigate the presence of the impact of the cost of business start-up procedures on socio-economic outcomes in the context of SSA economies. The results of our investigation allow for answering two broad questions of this study that we formulated at the beginning of our inquiry. The first question was:

1. *Whether ICT Capabilities impact CBSP?*

According to the collected evidence, we can answer this question as follows:

*There is evidence that high levels of ICT Capabilities in the areas of Affordability Readiness, Skills Readiness, Political & Regulatory Environment, and Business Usage are associated with low levels of CBSP.*

The second question was:

2. *Whether CBSP is associated with a socio-economic impact?*

The results of the data analysis offer sufficient evidence to answer the second question, as follows:

*The obtained evidence suggests that the socio-economic impact of CBSP is level-dependent, where low levels of the cost are associated with the presence of the impact, while high levels of the costs demonstrate no such association.*

By answering these questions, we are in a good position to address the overall question of our investigation, which is stated below:

*Whether ICT Capabilities can result in a positive socio-economic outcome by impacting the cost of business start-up procedures in the context of SSA economies?*

While based on the results of the data analysis the best short answer that we can provide to this question is "it depends", we can formulate a longer version of the answer, as follows:

*High levels of ICT capabilities, especially in the areas of Affordability Readiness, Skills Readiness, Political & Regulatory Environment, and Business Usage are associated with lower levels of CBSP, and, in turn, with a positive socio-economic outcome.*

We believe that this answer offers a valuable insight to policy and decision-makers, for it does link investments in ICT and the socio-economic outcomes of such investments by a transparent "Input → Engine → Output" process.

## CONCLUSION

There is a consensus that SSA economies represent a group of poorest economies in the world, and one of the challenges for the researchers and practitioners working in the area of *ICT for Development* (ICT4D) is to offer and implement interventions that will change the status quo. Fundamentally, the challenges are of two types: first, find effective paths by which investments are transformed into socio-economic outcomes, and, second, find the ways to optimize the paths- make the mechanisms of transformation of investments into outcomes more efficient.

Many SSA economies struggle with the problem of having a large number of informal SMEs and a low level of wealth, where formal SMEs contribute less than 20% to gross GDP versus around 60% in high-income countries (Fjose, Grünfeld, & Green, 2010). The link between the number of formal SMEs and the level of wealth of an economy, *ceteris paribus*, is apparent. The presence of this link is important, for it addresses the first challenge of finding an effective path of transformation of investments into economic outcomes via "investments → formalization of SMEs → socio-economic outcomes of investments" sequence of steps. And once the link is established, the optimization of the mechanism of transformation can take place.

Consequently, it seems reasonable to suggest that one of the ways of improving the macroeconomic bottom line of SSA economies is by increasing the share of formal SMEs. Our investigation dealt with one of the barriers to establishing a formal SME – the cost of business start-up procedures. We can claim, based on the results of the data analysis, that this study was a successful one – we were able to, first, identify the context-specific factors that impact legitimization of SMEs in the context of SSA economies, and, second, establish a link between the relative magnitude of the legitimization barrier (e.g., *CBSP*) and socio-economic impact.

One of the contributions of this investigation is a theoretical one – we have outlined a research framework allowing for investigating the socio-economic impact of *ICT Capabilities* via various engines of the impact. The suggested framework was not created from scratch; instead, it is an extension of the framework of NRI, and it is consistent with the grand theory of the Neoclassical Growth Accounting.

Another contribution is a methodological one – despite the fact that Samoilenko and Osei-Bryson (2017) suggested using CA instead of received categorization to identify heterogeneous sub-groups in a sample, to our knowledge this study is the first one to actually implement it as a part of a methodology in action.

We also believe that our study has some value for the practitioners working in the area of ICT4D – the results of our investigation offer some actionable insights, specifically in the form of "If → Then" rules, which could be put to practice in order to improve the socio-economic well-being of SSA economies.

We must acknowledge limitations of our inquiry. The first one is associated with the availability of the data – we wish we could include more SSA economies in our sample, but we were limited by what was offered by our data sources. The second limitation is a fairly narrow focus on a single factor- that of CBSP. While, undoubtedly, *CBSP* is important to consider, there are other possible factors that may serve as barriers to legitimization of SMEs. Nevertheless, we hope that insights offered by this investigation outweigh its limitation.

## ACKNOWLEDGMENT

Material in this chapter previously appeared in: Start a Business, Get A Credit, Make an Impact: Do ICTs Help? Empirical investigation in the Context of Sub-Saharan Economies. In *CONF-IRM* (p. 38).

## REFERENCES


Chipman, H. and Tibshirani, R. (2006). Hybrid Hierarchical Clustering with Applications to Microarray Data. Biostatistics Advance Access, 7(2), 286–301.

de Sousa dos Santos, J. F. (2015). Why SMEs are key to growth in Africa. World Economic Forum. Available on-line at: https://www.weforum.org/agenda/2015/08/why-smes-are-key-to-growth-in-africa/.

Di Battista, A., Dutta, S., Geiger, T., and Lanvin, B. (2015). The Networked Readiness Index 2015: Taking the Pulse of the ICT Revolution. Global Information Technology Report 2015, pp. 3–28.

Fierro, A. M. (2015). What are the biggest challenges for Africa's entrepreneurs? World Economic Forum. Available on-line at: https://www.weforum.org/agenda/2015/08/what-are-the-biggest-challenges-for-africas-entrepreneurs/.

Fjose, S., Grünfeld, L., and Green, C. (2010). SMEs and Growth in Sub-Saharan Africa: Identifying SME Roles and Obstacles to SME Growth, MENON-publication no. 14/2010. MENON Business Economics. Available on-line at: https://www.norfund.no/getfile.php/133983/Bilder/Publications/SME%20and%20growth%20MENON%20.pdf.

IFC. (2017). SME Initiatives. International Finance Corporation (World Bank Group). Available on-line at: http://www.ifc.org/wps/wcm/connect/REGION__EXT_Content/Regions/Sub-Saharan+Africa/Advisory+Services/SustainableBusiness/SME_Initiatives/.

IMF. (2015). Regional Economic Outlook: Sub-Saharan Africa. World Economic and Financial Surveys of International Monetary Fund. Available on-line at: http://www.imf.org/external/pubs/ft/reo/2015/afr/eng/.

Kirkman, S.G., Osorio, A.C., and Sachs, D.J. (2002). The Networked Readiness Index: Measuring the Preparedness of Nations for the Networked World. In The Global Information Technology Report 2001–2002 Readiness for the Networked World, Kirkman, G. (ed.). Oxford University Press, New York, NY, pp. 10–29.

Laan, M. and Pollard, K. (2002). A New Algorithm for Hybrid Hierarchical Clustering with Visualization and the Bootstrap. Journal of Statistical Planning and Inference, 117 (2), 275–303.

Samoilenko, S. and Osei-Bryson, K.M. (2008). Increasing the Discriminatory Power of DEA in the Presence of the Sample Heterogeneity with Cluster Analysis and Decision Trees. Expert Systems with Applications, 34(2), 1568–1581.

Samoilenko, S. and Osei-Bryson, K.M. (2017). A Methodology for Identifying Sources of Disparities in the Socio-Economic Impacts of ICT Capabilities in Sub-Saharan Economies. In Proceedings of the International Conference on Information Resource Management, Santiago, Chile, May 16–19, 2017.

Solow, R. (1957). Technical Change and the Aggregate Production Function. Review of Economics and Statistics, 39 (3), 312–320.

The World Bank (2016). Data Bank: World Development Indicators. Available on-line at: http://databank.worldbank.org/data/reports.aspx?source=world-development-indicators.

TWB (2016). Entrepreneurs and Small Businesses Spur Economic Growth and Create Jobs. The World Bank. Available on-line at: http://www.worldbank.org/en/news/feature/2016/06/20/entrepreneurs-and-small-businesses-spur-economic-growth-and-create-jobs.

WEF_NRI (2016). Networked Readiness Dataset. Available on-line at: http://reports.weforum.org/global-information-technology-report-2016/networked-readiness-index/.

# 24 Exploring the Socio-Economic Impacts of ICT-Enabled Public Value in Sub-Saharan Africa

## INTRODUCTION

Efficiency and effectiveness of a governmental apparatus are among the few qualities that are valued by citizens of any country. However, not every economy earns high marks in those areas (Kaufmann, Kraay, & Mastruzzi, 2010c); instead, most governments could use some improvement in the area of public services (Blaug et al., 2006). Consequently, interventions in this area (e.g., e-government initiatives) are common to most of the economies of the world. It was suggested that the practices and ideas of New Public Management (NPM) could serve as a tool for making administrative systems more reliable, consistent, and efficient (Hood & Lodge, 2006).

One of the ways of implementing NPM is by adopting information and communication technologies (ICTs) to create *ICT capabilities* that reduce cost and time of organizational activities (Cordella, 2006). However, the actual impact of ICT in this regard is not easy to assess because the formulations of NPM are quite diverse (Cordella & Bonina, 2012). Additionally, it was also noted that the commonly identified impacts of ICT on public administration in terms of economics, effectiveness, and efficiency of the provided services represent too narrow of a focus (Cordella & Bonina, 2012); thus, social impacts should also be taken into consideration.

Unsurprisingly, the relevant literature (Bannister & Connolly, 20141; Cordella & Willcocks, 2010) suggests that the notion of *public value* (Moore, 1995) is better suited to assess the impact of ICT on public initiatives (Cordella & Bonina, 2012). In this context public value can be defined as a benefit that is "…related to the achievements of objectives set by government programs and the delivery of public services to the citizens" (ibid, p. 516). It would appear then that the link "ICT Capabilities → Public Value" link is one that needs to be investigated.

However, a *Public Value* created as a result of ICT-enabled public sector initiatives is not an end in itself, but a means to an end – it was noted that investigating socio-economic outcomes of ICT-enabled public initiatives is of benefit and is preferable to inquiries that are more limited in scope (Cordella & Bonina, 2012). Furthermore, the question seems to be is not *if ICT Capabilities* allow for generating *Public Value* (e.g., its effectiveness), but how *efficiently* such value is generated (Jaeger, 2005; Eppler, 2007). Consequently, given the three constructs – *ICT*

*Capabilities, Public Value*, and socio-economic outcomes, we can state a general question of our inquiry, namely:

> *Do ICT Capabilities deliver positive socio-economic outcomes via impacting Public Value creation?*

This question is worth investigating in general (WEF_GITR, 2016), but it gains an additional importance in the context of the poorest countries of the world that we chose to be the subject of our inquiry – the economies of Sub-Saharan Africa (SSA). Three obvious reasons for why this is so come to mind. Firstly, results of the previous research indicated that SSA could boost their economic development via expanding their public sector (Kimaro, Keong, & Sea, 2017), and the expansion of the public sector could be positively impacted by ICT capabilities. This will necessitate the investments in ICT, for ICT capabilities cannot be created without the appropriate investments. In sum, SSA economies will have to invest in ICT.

Secondly, poorer economies have fewer resources to invest in ICT, and, as a result, they will be more dependent on efficiency of utilization of investments in ICT, as well as of efficiency of the impact of the resultant *ICT Capabilities*. Simply put, SSA economies are more vulnerable to the consequences of inefficiency of utilization of investments. Thirdly, it was noted that *Public Value* was found to influence the socio-economic development of SSA. There is evidence that bad governance negatively impacts economic performance (Habtamu, 2008); thus, it is only reasonable to expect that the reverse is also true and that *ICT Capabilities*-enabled "good governance" (e.g., generated *Public Value*) will have a positive impact on development of SSA economies.

Conceptually, the causal structure of the general question that we stated can be presented in the form of "ICT Capabilities → Public Value → Socio-Economic Outcomes". However, this simple deterministically causal chain of links is too simplistic for generating promising research questions – a more rigorous framework is needed. Unsurprisingly, the importance of developing theoretical underpinning linking ICT and public sector initiatives has been previously noted (Bekkers & Homburg, 2007; Madon, Sahay, & Sudan, 2007) and a possible mechanism linking the two (e.g., via efficient and effective information channels) has been acknowledged (Dunleavy et al., 2006; Gupta, Dasgupta, & Gupta, 2008). The first item on our agenda, therefore, is the placement of our inquiry on a solid theoretical footing that can support answering the general question. In the next part of the chapter we outline the research framework of this investigation to our readers.

## RESEARCH FRAMEWORK OF THE STUDY

In this investigation we rely on the framework of *Networked Readiness Index* (NRI), which serves as an established measure of ICT capabilities, or, *drivers*, of socio-economic development (WEF_GITR, 2015). The NRI framework is reflected by a single composite index (e.g., *NRI*) comprised of four sub-indexes – three *drivers* (*Environment*, *Readiness*, and *Usage* sub-indexes) and one *impact* sub-index.

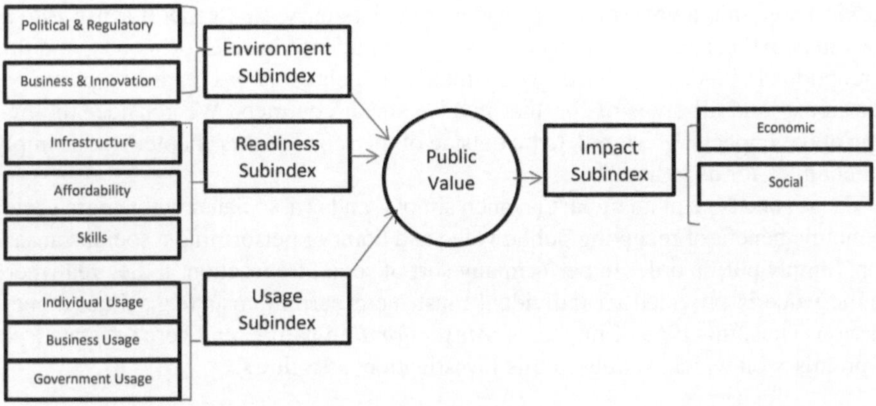


**FIGURE 24.1** The research framework of the study.

*Drivers* of the framework are not directly associated with *impact.* And it is because of this absence of the direct association between *drivers* and *impact* that the framework of NRI offers a great platform for exploring various factors (e.g., serving as an *engine of the impact*) that contribute to social and economic well-being of economies. For the purposes of our investigation we adapt the framework of NRI by using *public value* as the *engine of the impact* – the resultant framework is depicted in Figure 24.1.

This framework allows us to state two broad questions that we aim to answer within the context of SSA:

1. *How efficiently ICT Capabilities impact the creation of Public Value?*
2. *Whether the created Public Value is associated with a socio-economic impact?*

The conceptual underpinnings of these questions are simple. Let us consider the role of ICT and of the *ICT Capabilities* in any economy – their primary role is to create new information channels and/or to increase the efficiency and effectiveness of the existing ones. The role of ICT implementations in government and public services is no different in this regard (Eppler, 2007; Jaeger, 2005). By fulfilling this role ICT *should* reduce transaction costs and, as a result, improve the quality of public services and, consequently, generate public value. But does it do so effectively? Does it do so efficiently? We look closely at these questions in this study. To proceed further we need to clarify to our readers the *type of an impact* of *Public Value* that we aim to investigate.

There are at least two types of socio-economic impacts that *Public Value* can deliver. First one is associated with the *consequences of decisions* made by the citizens that receive public value as a result of effective and efficient functioning of their government. This type of an impact is a complex one, for it must account for a wide variety of benefits brought to all the beneficiaries of the received *Public Value* within the context where the value is provided. Let us consider *rule of law* as an instance of

*Public Value* that a government provides to its citizenry. In a general sense *rule of law* refers to the extent to which citizens have confidence in rules of society, and this perception of citizens will have a wide impact on their *decisions* regarding participation in any and all areas of life that involve social contracts. We consider the first type of the impact, *decisional*, to be outside of the scope of this chapter – it is simply too complex for us to handle.

The second type of an impact is much simpler and is associated with a pure socio-economic benefit of receiving public value at a point of performing a social transaction. Simply put, in order to perform any sort of social transaction in the area where public value is provided an individual must incur certain costs (e.g., time, money, energy). Thus, this type of impact is *transactional* in nature, and here we state a set of premises on which we rely in this investigation, as follows:

1. *Public Value provides a transactional benefit to the citizenry by impacting a socio-economic cost of social transactions, where greater Public Value results in lower socio-economic costs of social transactions.*
2. *Delivery and creation of Public Value is associated with effectiveness and efficiency of the agency tasked with the provision of the value.*
3. *ICT Capabilities can positively impact efficiency and effectiveness of the agency delivering public value via creation of new and optimization of the existing information channels.*

It is precisely this, *transactional*, type of the impact that we investigate in this study.

Let us now consider *Public Value* as an engine of socio-economic impact. First of all, it is only expected that a lower cost of performing a social transaction produces a direct economic impact, for if less money is allocated towards the transaction, then more money will be available for everything else, from business development to simple out-of-pocket spending. Similarly, we expect that a low social cost of transaction will produce a positive social impact, for less time and energy is allocated towards the transaction the more is left for other things and activities. But does, indeed, *Public Value* produce a socio-economic impact? We aim to answer this question in our inquiry.

Essentially, we approach the subject of this inquiry from the perspective of complex adaptive system, where the target system (e.g., a given SSA economy or group of economies) is comprised of three components (e.g., *ICT Capabilities, Public Value*, socio-economic outcomes) bound by non-linear relationships. And it is, fundamentally, the context-dependent strength of the relationships is of primary concern to this study. At this point we can formulate the overall question that we aim to answer, as follows:

> *Could ICT Capabilities result in a positive socio-economic outcome by impacting the creation of Public Value within the context of SSA economies?*

We conduct our investigation within the context of 26 economies of SSA, using the data set for the period of 2012–2016. The analysis of the data is supported by a five-phase methodology utilizing cluster analysis (CA), data envelopment analysis

(DEA), decision tree induction (DTI), association rule mining (ARM), and ordinary least squares (OLS).

## RESEARCH QUESTIONS OF THE STUDY

In our investigation we rely on the NRI framework (Kirkman, Osorio, & Sachs, 2002), the adapted version of which is depicted in Figure 24.2. An increase in the value of NRI for a given economy is indicative of the increase of the impact of ICT on innovation and productivity (Dutta & Jain, 2003). In this investigation we represent *Public Value* via the dimensions provided by *Worldwide Governance Indicators* (WGI) of the World Bank (The World Bank, 2016). The dimensions are *Political Stability and Absence of Violence/Terrorism*, *Voice and Accountability*, *Rule of Law*, *Regulatory Quality*, *Government Effectiveness*, and *Control of Corruption*.

It is only fair to our reader if we proactively deal with the elephant in the room that is an issue of *representation of Public Value*, and offer our answer to the question of "Why could the chosen dimensions serve as an adequate representation of *Public Value*?" It is worth noting that the issue of representation of any concept is a valid one to consider, for rarely there is a consensus regarding the choice of variables adequately describing a construct, let alone the scale of measurement of the variables. The issue is philosophical, as well as practical in nature, and it can never be completely settled due to the inherent limitation of modeling – the incompleteness of any representation.

However, what about the accuracy of representation? Do six *Worldwide Governance Indicators* accurately represent governance and, by extension, *Public Value*? In this chapter we take a position that the accuracy of representation of any non-trivial construct is fundamentally subjective and a consensus-based for a given domain. This is also, and may be even especially true, regarding the representation and measurement of governance, where the inherent difficulties of the undertaking

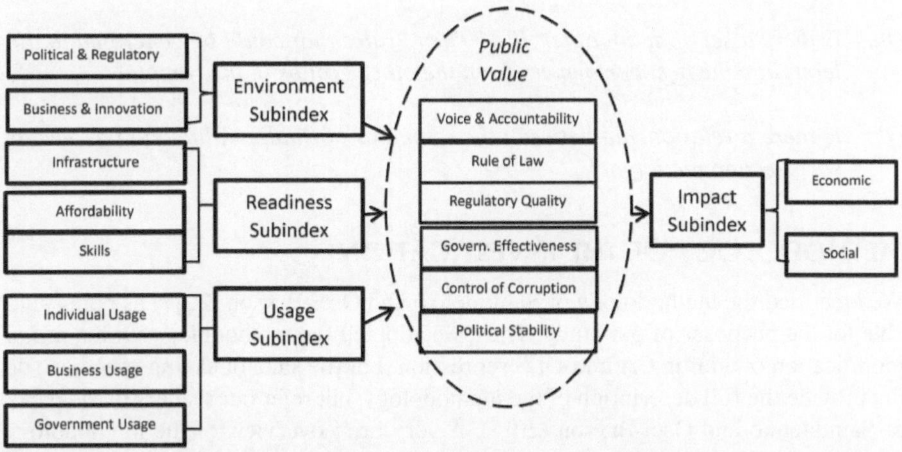


**FIGURE 24.2** The Unpacked Research Framework of this Study.

have been duly noted by Kaufmann, Kraay, and Mastruzzi (2010a). However, while construct validity of WGI is not immune to challenges (e.g., notably, Thomas, 2010), in the absence of better alternatives the definition of governance as it is supplied via WGI is perfectly acceptable for cross-country and over-time comparisons (Kaufmann et al., 2010b).

In summary, despite the marketplace for development of governance indicators being wide open (Kaufmann et al., 2007), WGI remains a long-standing, legitimate, and widely accepted tool for a representation of governance (Kaufmann, et al., 2010c). This may also be a good opportunity to present our reader with yet another proposition supporting this study, which links WGI, governance, and public value, as follows:

> *WGI can be used to represent governance, and one of the functions of governance is provision of Public Value to its citizenry.*

The representation of *Public Value* via WGI yields the framework depicted in Figure 24.2. Our reader will notice that the only difference between frameworks in Figures 24.1 and 24.2 is due to "unpacking" of public value into six indicators.

The research framework of this study allows for answering a set of important questions. We offer the formulation of the questions (*Q#:*) below.

*Q1:* *Do SSA economies in our sample represent a homogenous group in regard to NRI and Public Value?*

*Q2:* *If the sample of SSA economies is comprised of multiple sub-groups, then what differentiates the heterogeneous sub-groups?*

*Q3:* *In the case of the sample heterogeneity, how do sub-groups of SSA economies differ in terms of the relative efficiency of the conversion of ICT Capabilities into Public Value?*

*Q4:* *Is there a set of actionable "If → Then" rules that could be generated to differentiate the best performers from the other groups in the sample?*

*Q5:* *Is there a relationship between the estimates of Public Value and the level of socio-economic impact?*

## METHODOLOGY OF THE INVESTIGATION

We identified the methodology of Samoilenko and Osei-Bryson (2017) as being suitable for the purposes of our study. Thus, we adopted the methodology with a minor modification of adding CA and OLS regression. For the sake of saving space, we do not provide the full description of the methodology, but refer our readers to the paper of Samoilenko and Osei-Bryson (2017). A very brief overview of the methodology is provided below.

### Phase 1: Cluster Analysis

The purpose of CA is to identify naturally occurring sub-sets, or groupings, that may exist in the data set. The outcome of Phase 1 is N-cluster solution – partitioning of the sample into sub-groups is logical in nature (Samoilenko and Osei-Bryson, 2008), where no physical division of the sample takes place.

### Phase 2: Decision Tree Induction

DTI allows us to identify, based on the top-level splits, the variables that differentiate the clusters the most. The outcome of Phase 2 is a set of variables differentiating N-clusters comprising the sample.

### Phase 3: Data Envelopment Analysis (DEA)

The purpose of using DEA in this investigation is to evaluate the level efficiency with which ICT Capabilities transformed into public value. The outcome of Phase 3 is set of averaged scores for each of N clusters, which would allow us to identify the benchmark group in the sample of SSA economies.

### Phase 4: Ordinary Least Squares (OLS) Regression

The purpose of Phase 4 is to evaluate the presence of the causal relationships between public value and socio-economic outcomes. The outcome of Phase 4 is a set (21×N clusters) of *p*-values allowing for the testing of the significance of the relationships.

### Phase 5: Association Rule Mining (ARM)

ARM allows us to find possible patterns, associations, or causal structures that may exist in our data. The outcome of Phase 5 is a set of actionable "If → Then" rules based on the variables discovered in Phase 2.

## DATA

We obtained the data from two publicly available and highly reputable sources. NRI' data was downloaded from World Economic Forum' Networked Readiness Index' page (WEF_NRI, 2016) and the values of the variables representing *Public Value* were obtained from the Data Bank of the World Bank, Worldwide Governance Indicators database (The World Bank, 2016). In 2012 the number and representation of the *pillars* of three sub-indexes, the *drivers*, of NRI has changed, and it was also the year when the *impact* sub-index was introduced. Given the changes that took place between 2011 and 2012, we decided to concentrate on the new version of NRI and collect the data for the period between 2012 and 2016. Overall, we were able

**TABLE 24.1**
**Sample of Sub-Saharan Economies, by Income Level**

| Income Level | Sub-Saharan Economies |
|---|---|
| Low | Burundi, Chad, Ethiopia, Gambia, Kenya, Madagascar, Malawi, Mali, Mozambique, Rwanda, Tanzania, Uganda, Zimbabwe |
| Low-Middle | Cameroon, Cape Verde, Côte d'Ivoire, Ghana, Lesotho, Nigeria, Senegal, Swaziland, Zambia |
| Upper-Middle | Botswana, Mauritius, Namibia, South Africa |

to compile the data set representing 26 economies of SSA (the classification of the International Monetary Fund as of October 2014). Membership of each group of the sample is provided in Table 24.1.

## RESULTS OF THE DATA ANALYSIS

### Phase 1: CA

While performing CA we have followed a rule of thumb for the relative size of any single cluster being greater than 10% of the overall size of the sample (Samoilenko & Osei-Bryson, 2008). By performing hybrid hierarchical clustering (there seems to be a consensus that this approach combines the advantages of other methods (Laan & Pollard, 2002; Chipman & Tibshirani, 2006) without having any unique disadvantages), we arrived to 3-cluster solution. Results are summarized in Table 24.2.

We also calculated relevant average values of WGI per cluster. Results are presented in Table 24.3.

Based on the original solution 19 SSA economies are "permanent residents" of a given cluster, while 7 SSA economies are "migrants". This indicates that 77% of

**TABLE 24.2**
**Results of Clustering: Membership, Income Composition, and Size of Each Cluster**

| Cluster # | Composition | Cluster Membership |
|---|---|---|
| **Cluster 1**<br>Size:32 | LI 21.88%<br>LM 43.75%<br>UM 34.38% | Mauritius, South Africa, Cape Verde, Ghana, Zimbabwe (2013–2015), Botswana (2012), Uganda (2012–2015), Kenya (2014–2016) |
| **Cluster 2**<br>Size: 62 | LI 69.35%<br>LM 30.65%<br>UM 0.00% | Burundi, Chad, Ethiopia, Madagascar, Malawi, Mali, Mozambique, Tanzania, Swaziland, Côte d'Ivoire (2012–2014), Nigeria (2015–2016), Uganda (2016), Lesotho (2012–2015), Zimbabwe (2012, 2016) |
| **Cluster 3**<br>Size: 36 | LI 27.78%<br>LM 47.22%<br>UM 25.00% | Cameroon, Senegal, Zambia, Gambia, Rwanda, Namibia, Kenya (2012–2013), Côte d'Ivoire (2015–2016), Nigeria (2012–2013), Botswana (2013–2016) |

**TABLE 24.3**
**Comparison of Tree Clusters in Terms of the Values of WGI and Sub-Indexes of NRI**

| Cluster # | Voice and Accountability | Rule of Law | Regulatory Quality | Government Effectiveness | Control of Corruption | Political Stability | Average of WGI |
|---|---|---|---|---|---|---|---|
| **Cluster 1** | 0.28 | 0.04 | −0.04 | −0.03 | −0.16 | −0.03 | 0.01 |
| **Cluster 2** | −0.70 | −0.74 | −0.73 | −0.86 | −0.79 | −0.83 | −0.77 |
| **Cluster 3** | −0.31 | −0.24 | −0.17 | −0.32 | −0.11 | −0.08 | −0.20 |

| | Environment | Readiness | Usage | Impact |
|---|---|---|---|---|
| **Cluster 1** | 4.05 | 4.15 | 3.36 | 3.23 |
| **Cluster 2** | 3.30 | 2.69 | 2.64 | 2.66 |
| **Cluster 3** | 4.04 | 3.15 | 3.23 | 3.28 |

the clusters are stable in regard to their membership. Furthermore, if we consider four out of five years to be the indicator of stability, then over 92% of the clusters' membership is stable.

## Phase 2: DTI

By performing DTI, we were able to identify sub-indexes and pillars that differentiate the three clusters of SSA economies in our sample. During this phase we constructed various decision trees, based on different target variables (e.g., sub-indexes, pillars, NRI, and level of income). We were able to determine, for example, that the three clusters cannot be clearly differentiated based on the income level, or based on the value of NRI, or based on *impact* sub-index of NRI. However, we obtained evidence that the clusters do differ based on the values of *drivers* (sub-indexes), and the values of the associated pillars. We present summary of the most relevant results of DTI in Table 24.4.

**TABLE 24.4**
**Results of DTI: English Rules, Split Variables, and Split Criteria**

| English Rules | Split Variables |
|---|---|
| **Pillars-based DT** | **Pillars** |
| B04 > 3.880 | B04 – Readiness sub-index: Affordability |
| I B05 > 2.815: One {three=0, **one=32 (100%)**, two=0} | B05 – Readiness sub-index: Skills |
| I B05 ≤ 2.815: two {three=1, one=0, two=10} | A01 – Environment sub-index: Political and regulatory environment |
| B04 ≤ 3.880 | C07 – Usage sub-index: |
| I C07 > 3.220 | Business usage |
| I I A01 > 3.005: three {**three=35 (97%)**, one=0, two=5} | |
| I I A01 ≤ 3.005: two {three=0, one=0, two=4} | |
| I C07 ≤ 3.220: two {three=0, one=0, **two=43 (69.35%**} | |

## Phase 3: DEA

During the third phase we rely on DEA to evaluate relative efficiency of various "Drivers → public value" paths for each of the three clusters. We will use *input-oriented* model and variable *return to scale* (VRS) DEA model to conduct the analysis, for it is reasonable to argue that SSA economies have not yet reached the point of developing a level of ICT infrastructure allowing for accruing the benefits yielded by capitalizing on economies of scale. Description of the specific DEA models that we will be using in this phase is provided in Table 24.5.

Results of the DEA presented in Table 24.6. It is interesting to note that while there is practically no difference between the clusters in regard to two *Drivers → Impact* models, there is a significant difference between cluster 2 and the other two clusters in terms of *Drivers → Public Value* models. Simply put, cluster 2 is relatively less efficient in converting ICT capabilities into public value.

## Phase 4: OLS

In this phase of our methodology we run a regression analysis for each of the three clusters of SSA economies in our sample. For each cluster we create 21 OLS models, regressing the average of all six WGI', and each WGI, against, (1) Socio-economic impact (sub-index), (2) Social impact (pillar of *impact* sub-index), and (3) economic impact (pillar of *impact* sub-index). The results are summarized in Table 24.7.

**TABLE 24.5**
**DEA Model of the Study**

| DEA Model | Inputs of DEA Model | Outputs of DEA Model |
|---|---|---|
| *Drivers → Ave.Pub.Val.* | Environment, Readiness, Usage Sub-indexes | Average of 6 values of WGI |
| *Drivers → TotalPub.Val* | Environment, Readiness, Usage Sub-indexes | All six indicators of WGI |
| *Drivers → Voice & Accountability* | Environment, Readiness, Usage Sub-indexes | Voice & Accountability |
| *Drivers → Rule of Law* | Environment, Readiness, Usage Sub-indexes | Rule of Law |
| *Drivers → Regulatory Quality* | Environment, Readiness, Usage Sub-indexes | Regulatory Quality |
| *Drivers → Government Effectiveness* | Environment, Readiness, Usage Sub-indexes | Government Effectiveness |
| *Drivers → Control of Corruption* | Environment, Readiness, Usage Sub-indexes | Control of Corruption |
| *Drivers → Political Stability* | Environment, Readiness, Usage Sub-indexes | Political Stability |
| *Drivers → Overall Impact* | Environment, Readiness, Usage Sub-indexes | Impact Sub-index |
| *Drivers → Social + Economic Impact* | Environment, Readiness, Usage Sub-indexes | Social Impact Pillar<br>Economic Impact Pillar |

## TABLE 24.6
## Results of DEA

| DEA Model | Model Orientation | Relative Efficiency Scores, Averaged | | |
|---|---|---|---|---|
| | | **Cluster 1** | **Cluster 2** | **Cluster 3** |
| *Drivers → Public Value (average of WGI)* | *Input-oriented* | 0.96 | 0.91 | 0.97 |
| *Drivers → Public Value (all 6 of WGI)* | *Input-oriented* | 0.99 | 0.96 | 0.99 |
| *Drivers → Voice & Accountability* | *Input-oriented* | 0.95 | 0.90 | 0.97 |
| *Drivers → Rule of Law* | *Input-oriented* | 0.96 | 0.91 | 0.97 |
| *Drivers → Regulatory Quality* | *Input-oriented* | 0.96 | 0.90 | 0.98 |
| *Drivers → Government Effectiveness* | *Input-oriented* | 0.98 | 0.92 | 0.98 |
| *Drivers → Control of Corruption* | *Input-oriented* | 0.93 | 0.90 | 0.98 |
| *Drivers → Political Stability* | *Input-oriented* | 0.93 | 0.89 | 0.97 |
| *Drivers → Overall Impact* | *Input-oriented* | 0.93 | 0.94 | 0.97 |
| *Drivers → (Social Impact, Economic Impact)* | *Input-oriented* | 0.96 | 0.96 | 0.97 |

## TABLE 24.7
## Results of OLS

| OLS, Independent Variable | OLS, Dependent Variable | Significant at 5% Level? | | |
|---|---|---|---|---|
| | | **Cluster 1** | **Cluster 1** | **Cluster 1** |
| Average WGI | Economic Impact | **Yes** | No | **Yes** |
| | Social Impact | **Yes** | **Yes** | No |
| Voice & Accountability | Economic Impact | **Yes** | **Yes** | **Yes** |
| | Social Impact | **Yes** | **Yes** | **Yes** |
| Rule of Law | Economic Impact | **Yes** | No | **Yes** |
| | Social Impact | **Yes** | **Yes** | No |
| Regulatory Quality | Economic Impact | **Yes** | No | No |
| | Social Impact | **Yes** | No | No |
| Government Effectiveness | Economic Impact | **Yes** | No | **Yes** |
| | Social Impact | **Yes** | **Yes** | No |
| Control of Corruption | Economic Impact | **Yes** | No | **Yes** |
| | Social Impact | **Yes** | **Yes** | No |
| Political Stability | Economic Impact | No | No | **Yes** |
| | Social Impact | No | No | No |
| Overall significant (out of 6) | Economic Impact | 83% | 16% | 83% |
| | Social Impact | 83% | 67% | 16% |
| Effectiveness (average) | (Economic + Social Impact)/2 | 83% | 41% | 50% |

## Phase 5: ARM

We used *R* to run ARM analysis (the threshold was set at 20% for support and 60% for confidence).

Overall, we ran ARM analysis 19 times. The first time we analyzed the complete sample (all three clusters) for the presence of general "If → Then" rules that would involve any of the indicators of WGI on the right ("Then") side. After that we ran ARM six times, for each cluster, to identify rules that are specific to each of the indicators of WGI. The summarized results are provided in Table 24.8, while a larger set of chosen rules is provided in Appendix of the chapter. Looking at the rules provided in Table 24.8 our reader can easily notice two things. First, there seem to be an inverse relationship between ICT capabilities (as it is represented by the pillars of NRI) and public value (as represented by WGI). Second, there is a significant number of the association rules that contain *Political and Regulatory Environment* and *Business Usage* on the left side of "If → Then" equation – these are the variables that differentiate SSA economies in our sample.

**TABLE 24.8**
**Results of ARM, Summary**

| # | Left Side (*If*) | → | Right Side (*Then*) | Sup. | Conf. | Lift |
|---|---|---|---|---|---|---|
| Complete Set | HIGH Political & Regulatory Envir | → | LOW Reg. Quality | 0.21 | 0.82 | 2.98 |
| | HIGH Political & Regulatory Envir | → | LOW Rule of Law | 0.20 | 0.76 | 2.98 |
| | HIGH Skills Readiness | → | LOW Government Effect | 0.20 | 0.75 | 2.98 |
| | HIGH Infrastructure Readiness | → | LOW Government Effect | 0.20 | 0.62 | 2.38 |
| | HIGH Political & Regulatory Envir | → | LOW Control of Corruption | 0.20 | 0.67 | 2.63 |
| | LOW Political & Regulatory Envir | → | HIGH Rule of Law | 0.20 | 0.76 | 2.98 |
| | LOW Skills Readiness | → | HIGH Political Stability | 0.20 | 0.6 | 2.09 |
| | LOW Political & Regulatory Envir | → | HIGH Government Effect | 0.20 | 0.75 | 2.98 |
| | LOW Political & Regulatory Envir | → | HIGH Control of Corruption | 0.20 | 0.64 | 2.51 |
| **Control of Corruption** | | | | | | |
| 1 | HIGH Political & Regulatory Envir | → | MidLo Control of Corruption | 0.26 | 1.00 | 2.11 |
| 2 | LOW Political & Regulatory Envir | → | HIGH Control of Corruption | 0.30 | 0.64 | 2.00 |
| 3 | HIGH Political & Regulatory Envir<br>HIGH Infrastructure Readiness | → | LOW Control of Corruption | 0.28 | 1.00 | 2.25 |

| # | Left Side (*If*) | → | Right Side (*Then*) | Sup. | Conf. | Lift |
|---|---|---|---|---|---|---|
| **Government Effectiveness** | | | | | | |
| 1 | HIGH Infrastructure Readiness | → | LOW Government Effectiv | 0.47 | 0.83 | 1.40 |
| 2 | LOW Political & Regulatory Envir<br>LOW Business & Innovation Envir | → | HIGH Government Effectiv | 0.21 | 1.0 | 2.39 |
| 3 | HIGH Infrastructure Readiness | → | LOW Government Effectiv | 0.28 | 0.83 | 2.14 |
| **Political Stability** | | | | | | |
| 1 | HIGH Affordability Readiness<br>HIGH Skills Readiness<br>HIGH Government Usage | → | LOW Political Stability | 0.25 | 0.8 | 1.97 |
| 2 | LOW Affordability Readiness<br>LOW Business Usage | → | HIGH Political Stability | 0.21 | 0.72 | 1.91 |
| 3 | HIGH Political & Regulatory Envir<br>HIGH Individual Usage | → | LOW Political Stability | 0.22 | 0.89 | 2.29 |
| **Rule of Law** | | | | | | |
| 1 | HIGH Political & Regulatory Envir | → | LOW Rule of Law | 0.40 | 1.0 | 1.53 |
| 2 | LOW Political & Regulatory Envir LOW Business & Innovation Envir | → | HIGH Rule of Law | 0.21 | 1.00 | 2.39 |
| 3 | HIGH Political & Regulatory Envir HIGH Infrastructure Readiness | → | LOW Rule of Law | 0.25 | 0.90 | 2.70 |
| **Regulatory Quality** | | | | | | |
| 1 | HIGH Political & Regulatory Envir HIGH Business & Innovation Envir | → | LOW Regulatory Quality | 0.37 | 1.01 | 1.88 |
| 2 | LOW Business Usage | → | HIGH Regulatory Quality | 0.21 | 0.93 | 2.38 |
| 3 | HIGH Political & Regulatory Envir<br>HIGH Individual Usage | → | LOW Regulatory Quality | 0.22 | 0.89 | 2.00 |
| **Voice & Accountability** | | | | | | |
| 1 | HIGH Political & Regulatory Envir | → | LOW Voice & Account | 0.40 | 1.0 | 1.53 |
| 2 | LOW Business Usage | → | HIGH Voice & Account | 0.22 | 0.93 | 2.00 |
| 3 | HIGH Political & Regulatory Envir HIGH Individual Usage | → | LOW Voice & Account | 0.22 | 0.89 | 2.67 |

## DISCUSSION OF THE RESULTS OF THE STUDY

Prior to conducting inquiry, we identified five research questions that served as a set of stepping stones leading to answering two main questions of the study. At this point we are well-equipped to provide the answers (*A#*) to each of the research questions, below.

*A1: SSE economies do not constitute a homogeneous group in regard to their levels of ICT capabilities.*

*A2: The dimensions differentiating heterogeneous groups of SSA are represented by four pillars of NRI: Affordability Readiness, Skills Readiness, Political and Regulatory Environment, and Business Usage.*

*A3: The heterogeneous groups of SSA economies also differ in terms of the relative efficiency of the impact of ICT capabilities on public value, where the groups with higher levels of ICT capabilities and higher levels of WGI are relatively more efficient than the group with a lower level of ICT capabilities and a lower level of WGI.*

*A4: There is also a set of "If → Then" rules that differentiates the groups of SSA economies – in general high values of pillars of NRI are associated with low values of WGI, and vice versa.*

*A5: The obtained evidence suggests that the socio-economic impact of Public Value is level-dependent, and SSA economies with greater values of WGI are more effective in obtaining a socio-economic impact then the economies with the lower levels of WGI.*

The results of the data analysis allow for answering two broad questions of this study that we formulated in the beginning of our inquiry; the answers are as follows:

1. *There is evidence that high levels of ICT Capabilities in the areas of Affordability Readiness, Skills Readiness, Political and Regulatory Environment, and Business Usage allow for relatively more efficient generation of Public Value.*
2. *The obtained evidence suggests that the socio-economic impact of Public Value is level-dependent, where economies with higher levels of WGI are more effective in producing the impact.*

By answering these questions, we are in a good position to address the overall question of our investigation, as follows:

*High levels of ICT capabilities, especially in the areas of Affordability Readiness, Skills Readiness, Political and Regulatory Environment, and Business Usage are associated with the higher level of public value, and, in turn, with a positive socio-economic outcome.*

We believe that this answer offers a valuable insight to policy and decision-makers, for it does link investments in ICT and the socio-economic outcomes of such investments by a very logical and transparent "Input → Engine → Output" process.

Over the course of our inquiry we obtained sufficient evidence to support the existence of the hypothesized conceptual link "ICT Capabilities → Public Value → Socio-Economic Outcomes" – our mission seemed to be accomplished. However, because we also discovered the presence of somewhat counterintuitive association rules of the type "High ICT Capabilities → Low Public Value" and "Low ICT Capabilities → High Public Value", we would like to reconcile this finding with the overall results of the study.

The key to an explanation, in our view, can be found in the metadata of WGI. Specifically, WGI relies "exclusively on perceptions-based governance data sources" (Kaufmann et al., 2010c, p. 5) which "include surveys of firms and households, as well as the subjective assessments of a variety of commercial business information providers, non-governmental organizations, and a number of multilateral organizations and other public-sector bodies" (ibid., p. 5). This is consistent with the perspective on public value being defined by the consumers of the value and not the producers of the value (Alford & Hughes, 2008; Cordella & Bonina, 2012).

Keeping this in mind, we would like to offer to our readers a set of propositions on which we will base our rationalization for the presence of seemingly counterintuitive association rules. The propositions, presented in the form of a simple list, are as follows:

1. Increase in *ICT Capabilities* brings about an increase in ICT-related sophistication of the population.
2. A higher level of ICT sophistication of the population brings about a higher level of scrutiny of ICT-enabled products and services (e.g., government' public services) offered to the population.
3. A lower level of ICT sophistication of the population entails a lower level of scrutiny of ICT-enabled products and services offered to the population.
4. Resultantly, a population with a higher degree of ICT sophistication will apply, *ceteris paribus*, stricter criteria to appraise ICT-enabled products and services then a population with a lower degree of ICT sophistication.
5. The use of stricter assessment criteria will result, *ceteris paribus*, in lower assessment scores, and the use of softer assessment criteria will result in higher assessment scores.

The presented above five points are easy to follow and to critique – we invite our readers to do so. However, in our view the points above do explain why SSA economies with higher levels of *ICT Capabilities* appraise *Public Value* as being lower in quality as compared to their counterparts with lower levels of *ICT Capabilities*. This explanation is an intuitive one – let us consider, for example, a car mechanic, a person who is very familiar with car's engine and drivetrain. When purchasing a car, the mechanic's assessment of a vehicle will be more stringent, and will be performed by using harsher criteria, than the evaluation of a vehicle by a casual user. Simply put a common expression "familiarity breeds contempt", or as in the case of this study, "familiarity with ICT breeds contempt for ICT-enabled services" may not be off-mark.

## CONCLUSION

At this point there seem to be a tendency (Cordella & Bonina, 2012) to assess the impact of ICT initiatives in public sector in a fairly narrow, efficiency-based way (Kelly, Mulgan, & Muers, 2002). Unsurprisingly, it was noted that ICT-enabled initiatives in public sector should be evaluated on a broader scale than that of pure efficiency and effectiveness of the target processes (Bannister, 2007; Cordella & Bonina, 2012). In our investigation we expanded the horizon of the impact of ICT capabilities on public services to include socio-economic outcomes, and this we consider to be one of the contributions of our inquiry, for we obtained empirical evidence substantiating the conceptualized chain of links connecting *ICT Capabilities, Public Value*, and *Socio-Economic Outcomes*.

It has also been shown that the results of ICT-driven interventions could go beyond straight forward improvements in efficiency and effectiveness of the target processes and, instead (or, additionally, for Business Process Reengineering (transformative change) and Total Quality Management (incremental change) are not mutually exclusive), bring about transformative, qualitative changes. And, expectedly, the importance to consider ICT interventions in public service as carrying "potential consequences of the transformation of the relationship between the citizens and the state" has been duly noted (Cordella & Bonina, 2012, p. 515). Indeed, the results of our investigation suggest that ICT can have a transformative impact on the relationship between citizens and the creator of the public value. It seems that the growth in *ICT Capabilities* tend to bring about the re-assessment of public value generated by public services. Our findings suggest that while citizenry of the economies with low levels of *ICT Capabilities* may generously appraise the public services as being of high quality, this appraisal will turn more critical as a result of a greater level of sophistication of the populace that was brought about by increases in *ICT Capabilities*.

We consider the framework of this investigation to be another contribution of our effort.

Finally, we can claim a contribution to practice in the area of IT4D, for we obtained actionable insights regarding the existence of the condition-dependent relationships between *ICT Capabilities*, ICT-enabled public value, and socio-economic outcomes.

## ACKNOWLEDGMENT

Material in this chapter previously appeared in: Representation matters: An exploration of the socio-economic impacts of ICT-enabled public value in the context of sub-Saharan economies. *International Journal of Information Management*, *49*, 69–85 (2019).

## REFERENCES

Alford, J. and Hughes, O. (2008). Public Value Pragmatism as the Next Phase of Public Management. American Review of Public Administration, 38(2), 130–148.

Bannister, F. (2007). The Curse of the Benchmark: An Assessment of the Validity and Value of e-Government Comparisons. International Review of Administrative Sciences, 3(2), 612–631.

Bannister, F. and Connolly, R. (2014). ICT, Public Values and Transformative Government: A Framework and Programme for Research. Government Information Quarterly, 31(1), 119–128.

Bekkers, V. and Homburg, V. (2007). The Myths of E-Government: Looking Beyond the Assumptions of New and Better Government. !e Information Society, 23(5), 373–382.

Blaug, R., Horner, L., and Lekhi, R. (2006). Public Value, Politics and Public Management: A Literature Review. Work Foundation, London.

Chipman, H. and Tibshirani, R. (2006). Hybrid Hierarchical Clustering with Applications to Microarray Data. Biostatistics Advance Access, 7(2), 286–301.

Cordelia, A. (2006). Transaction Costs and Information Systems: Does IT Add Up?. Journal of Information Technology, 21(3), 195–202.

Cordelia, A. (2007). E-government: Towards the e-Bureaucratic Form?. Journal of Information Technology, 22(3), 265–274.

Cordella, A. and Willcocks, L. (2010). Outsourcing, Bureaucracy and Public Value: Reappraising the Notion of the "Contract State". Government Information Quarterly, 27(1), 82–88.

Cordella, A. and Bonina, C. (2012). A Public Value Perspective for ICT Enabled Public Sector Reforms: A Theoretical Reflection. Government Information Quarterly, 29(4), October 2012, 512–520.

Dutta, S., and Jain, A. (2003). The Networked Readiness of Nations. In The Global Information Technology Report 2002–2003, Dutta, S., A. Lanvin, B., & Paua, F. (eds.). Oxford University Press, New York, NY, Oxford.

Eppler, M. J. (2007). Information Quality in Electronic Government: Toward the Systematic Management of High-Quality Information in Electronic Government-to-Citizen Relationships. In Governance and Information Technology: From Electronic Government to Information Government, Mayer-Schönberger, V. and Lazer, D. (eds.), The MIT Press, Cambridge, MA.

Dunleavy, P., Margetts, H., Tinkler, J., and Bastow, S. (2006). Digital Era Governance: IT Corporations, the State, and e-Government. Oxford University Press, Oxford, UK.

Gupta, B., Dasgupta, S., and Gupta, A. (2008). Adoption of ICT in a Government Organization in a Developing Country: An Empirical Study. The Journal of Strategic Information Systems, 17(2), 140–154.

Habtamu, F. N. (2008). Roles of Governance in Explaining Economic Growth in Sub-Saharan Africa. Africa Policy Journal, 4, 1–21.

Hood, C. and Lodge, M. (2006). The Politics of Public Service Bargains. Oxford University Press, Oxford.

Jaeger, P. (2005). Deliberative Democracy and the Conceptual Foundations of Electronic Government. Government Information Quarterly, 22(4), 702–719.

Kaufmann, D., Kraay, A., and Mastruzzi, M. (2007). Answering the Critics. World Bank Policy Research Department Working Paper No. 4149. Washington, DC.

Kaufmann, D., Kraay, A., and Mastruzzi, M. (2010a). The Worldwide Governance Indicators: a Summary of Methodology, Data and Analytical Issues. World Bank Policy Research Working Paper no. 5431. Available at SSRN: http://ssrn.com/abstract=1682130.

Kaufmann, D., Kraay, A., and Mastruzzi, M. (2010b). Response to "What do the Worldwide Governance Indicators Measure?" The European Journal of Development Research, 22(1), 55–58.

Kaufmann, D., Kraay, A., and Mastruzzi, M. (2010c). The Worldwide Governance Indicators: Methodology and Analytical Issues. World Bank Policy Research Working Paper 5430. Available on-line at: https://www.brookings.edu/wp-content/uploads/2016/06/09_wgi_kaufmann.pdf.

Kelly, G., Mulgan, G., and Muers, S. (2002). Creating Public Value: An Analytical Framework for Public Service Reform. Strategy Unit of Cabinet Office, London, UK.

Kimaro, E. L., Keong, C. C., and Sea, L. L. (2017). Government Expenditure, Efficiency and Economic Growth: A Panel Analysis of Sub-Saharan African Low Income Countries. African Journal of Economic Review, 5(2), 34–54.

Kirkman, S.G., Osorio, A.C., and Sachs, D.J. (2002). The Networked Readiness Index: Measuring the Preparedness of Nations for the Networked World. In The Global Information Technology Report 2001–2002 Readiness for the Networked World, Kirkman, G. (ed.), Oxford University Press, New York, NY, pp. 10–29.

Laan, M. and Pollard, K. (2002). A New Algorithm for Hybrid Hierarchical Clustering with Visualization and the Bootstrap. Journal of Statistical Planning and Inference, 117 (2), 275–303.

Madon, S., Sahay, S., and Sudan, R. (2007). E-Government Policy and Health Information Systems Implementation in Andhra Pradesh, India: Need for Articulation of Linkages Between the Macro and the Micro. The Information Society, 23(5), 327–344.

Moore, M. (1995). Creating Public Value: Strategic Management in Government. Harvard University Press, Cambridge, MA.

Samoilenko, S. and Osei-Bryson, K.M. (2008). Increasing the Discriminatory Power of DEA in the Presence of the Sample Heterogeneity with Cluster Analysis and Decision Trees. Expert Systems with Applications, 34(2), 1568–1581.

Samoilenko, S. and Osei-Bryson, K.M. (2017). A Methodology for Identifying Sources of Disparities in the Socio-Economic Impacts of ICT Capabilities in Sub-Saharan Economies. In Proceedings of the International Conference on Information Resource Management, Santiago, Chile, May 16–19, 2017.

The World Bank (2016). DataBank: Worldwide Governance Indicators. Available on-line at: http://databank.worldbank.org/data/reports.aspx?source=worldwide-governance-indicators.

Thomas, M. A. (2010). What Do the Worldwide Governance Indicators Measure? European Journal of Development Research, 22, 31–54.

WEF_GITR. (2015). World Economic Forum' Global Information Technology Report. Available on-line at: http://reports.weforum.org/global-information-technology-report-2015/network-readiness-index/.

WEF_NRI (2016). Networked Readiness Dataset. Available on-line at: http://reports.weforum.org/global-information-technology-report-2016/networked-readiness-index/.

# 25 Contributing Factors to Information Technology Investment Utilization in Transition Economies

## *An Empirical Investigation*

### INTRODUCTION

What are the transition economies (TEs) and where, in the economic map of the world, do they fit? It is not an easy task to categorize the economies of the world according to the single taxonomy; for no single commonly agreed upon taxonomy exists. Perhaps, multiple perspectives shall do a better job of outlining the context of this study.

The World Bank classified all economies of the world based on the *gross national product (GNP) per capita* of a given country in 1999. If the value was less than \$755, the country was considered a *low-income* country. If the value was above \$9,266, then the country was labeled as a *high-income* country, and if the value falls somewhere in between, then it is a *middle-income* country (World Bank 2002). Middle-income countries can be further sub-divided into two sub-groups of *lower-middle-income* countries, where GNP per capita is between \$756 and \$2,995, and *upper-middle-income* countries, where GNP per capita is between \$2,996 and \$9,265.

The United Nations designates those low-income economies that are characterized by weak human assets (as measured through a composite *Human Assets Index*) and economic vulnerability as *least-developed* countries (UNCTAD, 2004) while conceding that no agreed upon criteria exists for categorizing either *developed* or *developing* countries (UNSD, 1999). According to the perspective of the World Bank, however, developed countries are the high-income economies that have a high standard of living (often represented in terms of *Human Development Index*; UNDP, 1998), while developing countries are the low- and middle-income economies that have a low to moderate standard of living.

The term *transition economies* is used to refer to countries in the process of transitioning from a government or a state-controlled centrally planned economy (Ollman, 1997; Myers, 2004) to a market-oriented economy, where the market, rather than government or state, plays the "invisible hand" (Smith, 1776). It does not mean, however, that TEs constitute a homogenous group in terms of the level of economic development. The World Bank, for example, may group some of them with the developed and some with the developing countries, depending on the level of industrialization.

Hoskisson et al. (2000) combined two groups of "51 high-growth developing countries in Asia, Latin America, and Africa/Middle East, and 13 transition economies in the former Soviet Union" into the category of *emerging economies.* They define "emerging economy" as a country that "satisfies two criteria: a rapid pace of economic development, and government policies favoring economic liberalization and the adoption of a free market system". They warn, "at present, there is no standard list of countries agreed to be emerging economies", nor that there exists a common agreement on the meaning of the term; consequently, the "term 'emerging market economy' may also mean different things to different researchers".

At this point, there is no single accepted taxonomy to classify economies of the world. There are differences between the countries that defy a single framework; nevertheless, there are common traits among them as well. One such trait is that the great majority of the countries in the world invest in *information and communication technologies* (*ICT*).

Revenue generation serves as a major means by which investments in ICT contribute to macroeconomic growth (UN ICT Task Force Report, 2005; WT/ICT Development Report, 2006). Consequently, improving the effectiveness and efficiency of revenue production is a possible route to increase the macroeconomic impact of investments in ICT. And while these investments have been consistently contributing to economic growth by producing significant and reliable streams of revenues in developed countries, the result of such investments is not as clear-cut in the context of the TEs. In order to make investments in ICT attractive to domestic and international investors, however, TEs must be able to demonstrate their ability to produce revenues from such investments in a reliable and efficient manner.

There is little doubt that investments in ICT could and do produce robust returns and contribute to the overall economic growth in the context of developed economies (OECD, 2005; Jorgenson, 2001; Jorgenson & Stiroh, 2000; Oliner & Sichel, 2002; Stiroh, 2002; Colecchia & Schreyer, 2001; Van Ark et al., 2002; Daveri, 2002; Jalava & Pohjola, 2002). In the contexts of developing countries and economies in transition, however, the levels of the returns on investments differ significantly.

The TEs present a particularly interesting case for the research because they share economic characteristics with both developed and less developed economic regions (OECD, 2004). For example, domestic markets of TEs "include *both* substantial populations with significant disposable income *and* large numbers of people without" (ibid, p. 12). As a result, TEs present a good vantage point from which the relationship between the investments in ICT and economic growth can be investigated.

Multiple research studies conducted in this area have identified a group of factors that affect the return on investments in ICT. It has been suggested that the differences in capital stock and infrastructure (Dewan & Kraemer, 2000; Piatkowski, 2002), a level of investments in ICT (Murakami, 1997; Piatkowski, 2002), as well as the amount and quality of the available human capital (OECD, 2004), are some of the variables that impact the level of returns on investments in ICT.

We point the attention of the reader to another, rather obvious, but overlooked, determinant of effective production of revenue, namely, the efficiency of utilization of investments. As mentioned above, there is a consensus in the research community that capital stock and infrastructure, human capital, and level of investments in ICT are factors that affect the economic outcome of investments. However, do these factors also affect the efficiency of utilization of investments? It would appear to be the case in the context of developed countries, but, to our knowledge, there have been no studies undertaken to answer this question in the context of TEs.

Furthermore, as economies gradually acquire better infrastructures, invest more in ICT, and improve the level of human capital, do these economies obtain higher levels of returns on investments because of the corresponding gradual increases in efficiency as well? We could not find any reported evidence in the published literature regarding this question either.

Let us elaborate on the importance of this research problem a little further. It is reasonable to propose that investments in ICT must be gauged in accordance with the level of efficiency with which these investments can be utilized. The simple supporting reason for this statement is that efficiency is a relative term. It is relative to the external context (i.e., how efficient is country A relative to country B?), as well as to the internal context, and it is the internal context that deals with the level of the available resources. If country A can efficiently utilize X level of the investments in ICT, in the absence of perfect scalability, the efficiency would likely change if the level of investments in ICT drastically changes.

However, the issue of the relationship and the interplay between the level of investments in ICT and the efficiency of utilization of investments are simply too complex to be tackled in this study, for there are smaller questions that we must answer first. One such question concerns the factors that affect the level of efficiency of utilization of investments in ICT.

In line with related research in this area, this study too is ultimately concerned with the economic outcome of investments in ICT. Unlike the other studies, however, this research inquires specifically into the efficiency of the process by which investments in ICT are utilized, as well as into factors that possibly affect the level of efficiency.

More formally, in this research, we look at a subset of investments in ICT, specifically *investment in telecoms*, and investigate how efficiently TEs utilize these investments to produce revenues, and what factors contribute to the efficiency of investment utilization. For the purposes of our study, we adapt the definition of investment in telecoms provided by the Yearbook of Statistics (2004) (see more about this source in the next section), namely, as an investment that:

> *refers to expenditure associated with acquiring the ownership of telecommunication equipment infrastructure (including supporting land and buildings and intellectual and non-tangible property such as computer software). These include expenditure on initial installations and on additions to existing installations.*

In order to address this question, we employ a three-phase methodology utilizing *data envelopment analysis* (DEA) (Charnes, Cooper, Lewin, & Seiford, 1994), *cluster analysis* (CA) (Aldenderfer & Blashfield, 1984), and *decision trees* (DT) (Breiman, Friedman, Olshen, & Stone, 1984). First, we discuss the theoretical framework that supports our inquiry.

## THEORETICAL FRAMEWORK

### Growth Accounting

To approach our research problem, we rely on a neoclassical framework of growth accounting. This framework originated from the work of Solow (1957) and since then has been widely used by other researchers (Oliner & Sichel, 2002).

The objective of growth accounting is to decompose, using a neoclassical production function, the rate of growth of an economy into the contributions from the different inputs. A neoclassical production function relates output and inputs in the following manner:

$$Y = f(A, K, L)$$

where $Y$ = output (most often in the form of GDP), $A$ = the level of technology/total factor productivity (TFP), $K$ = capital stock, and $L$ = quantity of labor/size of labor force.

Which, in the case of this study, becomes:

$Y$ = revenues from telecoms; $A$ = TFP; $K$ = investments in telecoms; and $L$ = number of full-time telecom employees.

Based on the function provided above, growth accounting uses the Cobb–Douglas production function:

$$Y = A \times K^{\alpha} \times L^{\beta}$$

where $\alpha$ and $\beta$ are constants determined by technology.

In the case of constant returns to scale (e.g., if $\alpha + \beta > 1$, then returns are increasing to scale, and if $\alpha + \beta < 1$, then returns are decreasing to scale), $\alpha + \beta = 1$, thus, $\beta = 1 - \alpha$, which gives the following formulation:

$$Y = A \times K^{\alpha} \times L^{1-\alpha}$$

It is important to note that this function does not necessarily represent the true relationships between the inputs and the output; rather, its purpose is simply to serve as a vehicle for exploration and interpretation of the macroeconomic outcomes.

Out of three inputs used by growth accounting, only capital $K$ and labor $L$ could be observed in the data, while TFP would serve as a residual (often referred to as *Solow's residual*) term capturing that contribution to $Y$ (GDP or revenues from telecoms), which is left unexplained by the inputs of capital and labor.

In the case of this study, we are interested, first, in the efficiency of the utilization of the investments in telecoms by full-time telecom employees and, second, in some of the factors that possibly contribute to the level of efficiency. Consequently, for every point in time (for every year in our case) neoclassical production function allows us to relate investments in telecoms, full-time telecom employees, and revenues from telecoms in the following fashion:

Revenues from telecoms = *f* (TFP, investments in telecoms, full-time telecom employees)

As a result, for each TE in the study, for the period of, let us say, ten years we have ten values of the relative efficiency of utilization of investments in telecoms by full-time telecom staff. This approach of determining efficiency at a *point in time*, rather than determining the change in efficiency *over a period of time*, has two advantages. First, we do not need to account for depreciation of the telecom infrastructure, assuming the same rate of depreciation for all TEs in the study. Second, because the number of full-time telecom employees is reported annually, it allows us to treat the number of full-time telecom employees as being constant over the period of one year.

One of the appeals of using the neoclassical growth accounting framework lies in its simplicity; after all, only two factors, the TFP growth and the rate of increase in inputs, are used to explain the growth rate of the output. As a result, this relationship reflects the fundamental assumptions of the framework, namely, the presence of technological progress and the growth of labor. However, the flip side of this simplicity is the somewhat limited explanatory capability of economic growth. For example, while assuming technological progress the framework neither explains the sources of the progress or the factors that affect the progress nor does it account for any possible interactions between the technological progress and capital growth. In reality, though, capital investments would be affected by the technological progress, for progress in information technologies have fueled capital investments in the economies of the United States and other developed countries.

Finally, according to another assumption of the growth accounting framework, namely that the capital is subject to the law of diminishing returns, the convergence of the poor and wealthy economies must take place. The reality, however, reflects that the gap between poor and rich countries of the world is widening.

Nevertheless, the use of the growth accounting framework for the purposes of researching contributions of ICT investments to macroeconomic growth of TEs appears to be warranted, for this analytical framework has been widely used to estimate the contribution of ICT to economic growth in the context of developed and developing countries (Oliner & Sichel, 2000; Schreyer, 2000; Daveri, 2000; Jorgenson & Stiroh, 2000; Whelan, 1999; Hernando & Nunez, 2002).

Rarely, however, should any type of investment made on the macroeconomic level be perceived and considered in isolation. Rather, it is more beneficial to search and identify a set of complementary factors that allow magnifying the potential benefits of the investment. Often, such complementary factor, or a set of factors, is represented by other types of investments, sometimes in the entirely different areas of the economy. In this chapter, we use *theory of complementarity* as a theoretical framework that supports our search for the factors that may contribute to the efficiency of utilization of investments in telecoms by full-time telecom employees. A brief overview of the theory of complementarity is offered next.

## Theory of Complementarity

Initially introduced in economics by Edgeworth (1881), the concept of complementarity refers to the notion that the increase in one factor may result in the increased benefit received from its complementary factors. We apply the theory of complementarity to our research problem to argue that if the benefits of the investments in telecoms and in full-time telecom labor are to be reaped successfully at the macroeconomic level, then such investments should not be made in isolation from investments in other areas.

Thus, if two factors are more effective when taken jointly, rather than separately, we consider such factors complementary. We suggest that the process of identification of complementary factors requires a two-step approach. First, we need to identify a pool of possibly complementary factors. Second, we then proceed further and test those factors for complementarity with the variable of interest.

In the case of our study, we are interested in finding a set of factors that may contribute to the efficiency of the utilization of the investments in telecoms by full-time telecom employees. Consequently, our research can be perceived as aiming to accomplish the first part of the two-step approach. Namely, the aim of our study is to identify a pool of the factors that are possibly complementary to the number of full-time telecom employees.

However, even if the complementarity of the investments exists within a given production function, it cannot be identified through the formulation offered by Cobb–Douglas production function. Complementarity of the investments can only be discerned if the formulation allows for the presence of the interaction term between the specified investments. Thus, we turn our attention to the transcendental logarithmic production function, a brief overview of which is offered next.

Standard Cobb–Douglas production function, as it was mentioned before, can be represented as:

$$Y = A \times K^{\alpha} \times L^{\beta}$$

By taking the logarithm, the following formulation is obtained:

$$\log Y = \log A + \alpha \log K + \beta \log L$$

Extension to the above formulation of the Cobb–Douglas production function, called the transcendental logarithmic (*translog*) production function, is provided below:

$$\log Y = \beta_0 + \beta_1 \times \log K + \beta_2 \times \log L + \beta_3 \times \log K^2 + \beta_4 \times \log L^2 + \beta_5 \times \log K \times \log L + e$$

It is easy to see that the Cobb–Douglas function is "nested" in the translog function, and testing whether both functions describe the production process equally well would entail testing the following null hypothesis:

$$H_0 : \beta_3 = \beta_4 = \beta_5 = 0$$

The translog production function is more flexible than the Cobb–Douglas function in the sense that it allows testing for the presence of the interactions between the variables, where the test for the presence of the interaction would involve testing of the following hypothesis:

$$H_0 : \beta_5 \text{ is not statistically discernible from 0 at the given level of } \alpha$$

Complementarity of investments has been investigated in the context of research and development (R&D) portfolios by Lambertini (2003), Lin and Saggi (2002), Rosenkranz (2003) and in the context of process and product innovation by Athey and Schmutzler (1995). In a more relevant context to this research, Giuri, Torrisi, and Zinovyeva (2005) explored the complementarity between skills, organizational change, and investments in ICT. Bugamelli and Pagano (2004) studied the complementarity between investment in ICT and the related investment in human and organizational capital. Gera and Wulong (2004) examined the complementarity of the investment in ICT and organizational changes and worker skills, and Loukis and Sapounas (2004) inquired into the complementarity between IS investment and the set of IS management factors.

## OVERVIEW ON THE DATA

In our choice of data analytic tools, we were restricted by our selection of the data which represents a sample of convenience. Use of any parametric method would have required an assumption of data normality, which a sample of convenience may not satisfy. Thus, we decided to use DEA, CA, and DTs, all well-known and dependable non-parametric methods.

The data for this study were obtained from the *World Development Indicators* database (web.worldbank.org/WBSITE/EXTERNAL/DATASTATISTICS), which is the *World Bank*'s (web.worldbank.org) comprehensive database on development data, and the *Yearbook of Statistics* (2004) (www.itu.int/ITU-D/ict/publications), which is published yearly by *International Telecommunication Union (ITU)* (www.itu.int). In our choice of variables, we were greatly restricted by the availability of the data. For example, while the development data of the World Bank's database covers more than 600 indicators for 208 economies, data on many of the indicators relevant to our research were not available, or were available only for a few countries, or contained too few data points to be useful in data analysis.

In our choice of TEs to include in the study, we tried to identify and select a group of countries that started the transition process at approximately the same time. Thus we decided on the following 18 transitional economies: Albania, Armenia, Azerbaijan, Belarus, Bulgaria, Czech Republic, Estonia, Hungary, Kazakhstan, Kyrgyzstan, Latvia, Lithuania, Moldova, Poland, Romania, Slovakia, Slovenia, and Ukraine.

In terms of the length of the time series, we were restricted to the period from 1992 to 2004, for which data were provided by the Yearbook of Statistics of ITU. We decided to begin our analysis with the year 1993 because we believe that year provides a common starting point for the transitional economies. Our reasoning here is that it took a year from the dissolution of the Soviet bloc in 1991 for the transition process to begin, and using the year 1992 as a starting point may favor "early starters".

## METHODOLOGY: SEARCHING FOR THE DETERMINANTS OF THE EFFICIENCY OF UTILIZATION OF INVESTMENTS IN TELECOMS

In this part of the chapter, we describe, in step-by-step fashion, the methodology used in this study.

### PHASE 1: DATA ENVELOPMENT ANALYSIS

The cornerstone of our approach is DEA which we utilize to obtain the relative efficiency score for each TE in our sample. Our data set spans a 10-year period, consequently, we perform DEA ten times, ones for each year for the period from 1993 to 2002. As a result, we obtain ten scores of the relative efficiency, one for every year, for each TE in our sample. These scores refer to the relative efficiency of transforming investments in telecoms into revenues from telecoms. From the set of available data, we select a subset representing inputs, and a subset representing outputs. These are the variables that are used in the specification of the DEA model.

#### Data Used to Perform DEA

For the DEA part of the methodology, we have identified a model consisting of six input and four output variables. We present the description of the model first and then follow with the justification of the variables that comprise our model.

Input variables of the DEA model:

1. GDP per capita (in current US $)
2. Full-time telecommunication staff (% of the total labor force)
3. Annual telecom investment per telecom worker
4. Annual telecom investment (% of GDP in current US $)
5. Annual telecom investment per capita
6. Annual telecom investment per worker

Output variables of the DEA model:

1. Total telecom services revenue per telecom worker
2. Total telecom services revenue (% of GDP in current US $)
3. Total telecom services revenue per worker
4. Total telecom services revenue per capita

The main goal that we pursue in performing DEA is to find out how efficient the 18 transition economies are in converting investment inputs into the revenue outputs. Therefore, we did not include any other types of inputs or outputs such as those related to infrastructure, capabilities, and utilization. It is should be mentioned that the purpose of our DEA model is not to reflect the path by which the investments are transformed into the revenues over the course of one year, rather, the intent of our model is to depict a "fiscal efficiency" of the TEs regarding their investments in telecoms.

Upon the close inspection of the chosen variables, one can see that all of them are expressed as ratios. We intentionally present the levels of investments and revenues not in absolute dollar terms, but in relative units. The intent in doing so is to lessen the impact of the differences between TEs in terms of their size, population, level of wealth while representing the investments and revenues more broadly (i.e., relative to the whole population, labor force of a country, and the telecom industry). We argue that such relative representation provides a more objective depiction of not only the investments and revenues themselves but also of the economic and demographic environment within which the investments take place and the revenues produced.

There are no objective criteria according to which the "best" DEA model can be constructed; instead, the decision about including input and output variables is usually delegated to the purview of the investigator. We would like to, however, provide some justification regarding our choice of the input and output variables in our DEA model.

We include the input variable *GDP per capita (in current US $)* in order to take into consideration the differences between the levels of the economic development of eighteen TEs in the study. Let us recall, that according to the World Bank, economies can be classified as low-income, lower-middle-income, upper-middle-income, and high-income. Consequently, the inclusion of the input variable *GDP per capita (in current US $)* allows us to account for the possible differences in the level of industrialization of these countries.

The reason for the inclusion of the input variable *full-time telecommunication staff (% of the total labor force)* is intuitive; according to the assumption of the study, investments in telecoms are converted into revenues by full-time telecom employees, who represent one of the essential input components of the revenue-generating process (i.e., without employees, investments cannot be converted into revenues). The inclusion of the rest of the input and output variables of the DEA model is based on the theoretical framework used in our study, neoclassical growth accounting. Again, the reason for representing the variables *annual telecom investments* and *total telecom services revenues* relative to GDP, total population, total labor force, and total telecom employees, was to counter the differences between TEs in terms of their size, population, and level of wealth. At the same time, such an approach allows us to obtain some sort of representation of the structure of the economies within which investments convert to revenues.

## Phase 2: Cluster Analysis

In the second phase of our inquiry, we use CA to determine whether the TEs in our sample are similar in terms of their relevant characteristics, as represented by the input and output variables of the DEA model. Thus, more formally, in this part of the study, we test the null hypothesis that there are no discernable clusters of TEs with respect to their level of investments in and revenues from ICT. This hypothesis can be stated as follows:

> **$H_0$:** *The sample of 18 TEs is homogenous in terms of the levels of annual telecom investments and total telecom services revenues*

Consequently, we reject the null hypothesis if CA results in more than one cluster and, given a set of data points representing a transitional economy over a 10-year period of time, every cluster contains a complete set of data points representing a given economy.

If the results of CA reveal the presence of multiple subgroups in the sample, then we calculate the averaged over 10 year-period relative efficiency scores (identified in

Phase 1) for each group. We can expect that if heterogeneous subgroups are identified in our sample, then these subgroups have different average relative efficiencies.

### Data Used to Perform CA

To perform CA, we reduce the data set used to conduct DEA by removing the *GDP per capita* and *full-time telecommunication staff* variables. While these two variables are important as the inputs of the DEA model, in CA we aim to test homogeneity of the sample in terms of the investments and revenues only, thus, we decided against using these variables in CA. The complete list of the variables used to perform CA is provided below.

Variables used to perform CA:

1. Annual telecom investment per telecom worker (Current US $)
2. Annual telecom investment (% of GDP)
3. Annual telecom investment per capita (Current US $)
4. Annual telecom investment per worker (Current US $)
5. Total telecom services revenue per telecom worker (Current US $)
6. Total telecom services revenue (Current US $)
7. Total telecom services revenue per worker (Current US $)
8. Total telecom services revenue per capita (Current US $)

## Phase 3: Decision Tree

In the third phase of our study, we use DT analysis to identify the most important dimensions that differentiate the heterogeneous subgroups in our sample. The goal of this phase is to identify some of the variables that may be responsible for the differences in average relative efficiencies across the subgroups. In order to do so, we create a new categorical target variable with its domain of values equal to the number of subgroups identified in Phase 2. Once the first, most important split is made, we record the name of the variable and the value at which the split was made. After that, we remove that variable from further analysis and repeat the procedure.

### Data Used to Perform DT

Before conducting the DT analysis, we identified the largest set of data available to us; in our analysis, we were able to use 34 variables, which are listed below. Let us recall, that in this study we are interested in finding a set of factors that may contribute to the efficiency of the utilization of the investments in telecoms by full-time telecom employees. Consequently, the aim of this part of the data analysis is to identify a pool of factors that are possibly complementary to the number of full-time telecom employees.

Variable used for DT analysis:

1. Exports of computer, communications and other services (% of commercial service exports)
2. High-technology exports (% of manufactured exports)
3. Imports of computer, communications and other services (% of commercial service imports)

4. Military expenditure (% of GDP)
5. Military personnel (% of the total labor force)
6. Fixed line and mobile phone subscribers (per 1,000 people)
7. International telecom, outgoing traffic (minutes per subscriber)
8. Internet users (per 1,000 people)
9. Mobile phones (per 1,000 people)
10. Telephone mainlines (per 1,000 people)
11. Telephone mainlines per employee
12. Health expenditure per capita (current US $)
13. Health expenditure, private (% of GDP)
14. Health expenditure, public (% of GDP)
15. Health expenditure, total (% of GDP)
16. Immunization, DPT (% of children ages 12–23 months)
17. Immunization, measles (% of children ages 12–23 months)
18. Pupil-teacher ratio, primary
19. School enrollment, secondary (% gross)
20. School enrollment, tertiary (% gross)
21. Research and development expenditure (% of GDP)
22. Researchers in R&D (% of the total labor force)
23. Technicians in R&D (% of the total labor force)
24. Roads, paved (% of total roads)
25. Roads, total network (km)
26. Full-time telecommunication staff (% of the total labor force)
27. Annual telecom investment (% of GDP in current US $)
28. Urban population (% of total)
29. Urban population growth (annual %)
30. Population growth (annual %)
31. Foreign direct investment, net inflows (% of GDP)
32. GDP growth (annual %)
33. GDP per capita (constant 2000 US $)
34. GDP per capita growth (annual %)

# RESULTS

## RESULTS: DEA

In this section, we describe the results of the DEA. To perform DEA, we used the software application "OnFront", version 2.02, produced by Lund Corporation (www.emq.com).

In using "OnFront" to obtain the efficiency scores, we have chosen to use Farrel Input-Saving Measure of Efficiency as a direct efficiency measure for the three types of models: CRS (constant return to scale), VRS (variable return to scale), and NIRS (non-increasing return to scale). In Table 25.1, we provide the averaged over 10 years scores of the relative efficiency of the 18 TEs. The complete results of DEA, with the scores for each year, are provided in Appendix A.

The results of DEA show that a number of transitional economies in some years obtained a rating of hundred-percent relative efficiency. This, sometimes overly

**TABLE 25.1**
**Averaged Scores of Relative Efficiency, per Country, per DEA Model**

| Country | CRS | VRS | NIRS |
|---|---|---|---|
| Albania | 0.98 | 1.00 | 0.98 |
| Armenia | 0.89 | 0.91 | 0.89 |
| Azerbaijan | 0.73 | 0.97 | 0.73 |
| Belarus | 0.51 | 0.68 | 0.51 |
| Bulgaria | 0.92 | 0.94 | 0.94 |
| Czech Republic | 0.85 | 0.90 | 0.86 |
| Estonia | 0.96 | 0.97 | 0.96 |
| Hungary | 1.00 | 1.00 | 1.00 |
| Kazakhstan | 0.64 | 0.77 | 0.67 |
| Kyrgyzstan | 0.93 | 1.00 | 0.93 |
| Latvia | 0.76 | 0.89 | 0.76 |
| Lithuania | 0.66 | 0.83 | 0.67 |
| Moldova | 0.92 | 1.00 | 0.92 |
| Poland | 0.93 | 0.96 | 0.93 |
| Romania | 0.58 | 0.70 | 0.58 |
| Slovakia | 0.81 | 0.84 | 0.81 |
| Slovenia | 0.95 | 0.98 | 0.95 |
| Ukraine | 0.91 | 0.93 | 0.93 |

generous assignment of efficiency scores, is a common characteristic of the most DEA models (Lins et al., 2003). Another common characteristic of DEA models is that they tend to evaluate as efficient those DMUs that have the smallest input values, or, the DMUs with the largest outputs (Ali, 1994).

Consequently, based only on the results of the DEA analysis we cannot determine the true nature of the relative efficiency of the TEs in our sample. It is possible that the relatively efficient TEs obtained their status because they are indeed more efficient in the utilization of the inputs than the relatively inefficient ones. It is also possible, however, that the status of being relatively efficient was awarded to the TEs with the lowest levels of the investments in telecoms.

## Results: Cluster Analysis

Let us recall that we conducted CA with the purpose of identifying the presence of possible differences between the TEs in terms of the levels of investments and the revenues.

The variables subjected to CA are not measured on the same scale, so, prior to CA the data had to be standardized. We used SAS Enterprise Miner (EM) to perform CA. We started our inquiry by choosing "Automatic" setting, which did not require any input from the investigator regarding the desired number of clusters.

The "Automatic" setting of EM uses *Standard Least Squares* clustering criterion (which minimizes the sum of squared distances of data points from the cluster

**TABLE 25.2**
**Membership of the Two-Cluster Solution**

| Majority | Leaders |
|---|---|
| **Albania (1993–2002)** | Bulgaria (2002) |
| **Armenia (1993–2002)** | **Czech Rep (1993–2002)** |
| **Azerbaijan (1993–2002)** | Estonia (1994–2002) |
| **Belarus (1993–2002)** | **Hungary (1993–2002)** |
| Bulgaria (1993–2001) | Latvia (1994, 1995, 1997–2002) |
| Estonia (1993) | Lithuania (1999–2002) |
| **Kazakhstan (1993–2002)** | **Poland (1993–2002)** |
| **Kyrgyzstan (1993–2002)** | **Slovenia (1993–2002)** |
| Latvia (1993, 1996) | Slovakia(1995–1998, 2000–2002) |
| Lithuania (1993–1998) | |
| **Moldova (1993–2002)** | |
| **Romania (1993–2002)** | |
| Slovakia (1993, 1994, 1999) | |
| **Ukraine (1993–2002)** | |

means), *Ward's Minimum Variance*[1] as a clustering method, and limits the minimum number of clusters to 2 and the maximum to 40. By beginning with this setting, which resulted in a five-cluster solution, we were able to determine the starting point in our analysis. By requesting fewer and fewer number of clusters, we then gradually derived four-, three-, and two-cluster solutions.

By using CA, we were able to come up with a solution that partitions our data set into two clusters. The membership of each cluster is provided in Table 25.2. One of the clusters contains the data points completely representing Poland, Czech Republic, Hungary, and Slovenia over the 10-year period, while the second cluster contains the data points completely representing Albania, Armenia, Azerbaijan, Belarus, Kazakhstan, Kyrgyzstan, Moldova, Romania, and Ukraine. Thus, these results suggest that we are able to reject the null hypothesis regarding homogeneity of 18 TEs in terms of investments and revenues from telecoms.

Once the results of CA were obtained, we separated our data set into the two subgroups, and calculated the scores of the averaged relative efficiency for each cluster. According to these calculations, one of the clusters, members of which include Czech Republic, Hungary, Poland, and Slovenia, has higher averaged relative efficiency scores than the cluster containing Albania, Armenia, Azerbaijan, Belarus, Kazakhstan, Kyrgyzstan, Moldova, Romania, and Ukraine. Subsequently, we call the first group the "Leaders" and the second group the "Majority".

These findings are consistent with the results reported by Piatkowski (2003), who studied eight transition economies of Europe (Bulgaria, Czech Republic, Hungary, Poland, Romania, Russia, Slovakia, and Slovenia) and concluded that in the period

[1] *SAS System Documentation* offers following description: "Ward's method tends to join clusters with a small number of observations, and it is strongly biased toward producing clusters with roughly the same number of observations. It is also very sensitive to outliers (Milligan, 1980)".

**TABLE 25.3**
**Comparison of the Clusters Based on DEA**

| Criterion for Comparison | "Leaders" Cluster | "Majority" Cluster | Difference | Difference (%) |
|---|---|---|---|---|
| Average efficiency score, CRS | 0.89 | 0.79 | 0.10 | 12.54 |
| Average efficiency score, VRS | 0.95 | 0.88 | 0.07 | 7.48 |
| Average efficiency score, NIRS | 0.89 | 0.80 | 0.09 | 11.63 |

between 1995 and 2000 ICT capital has most potently contributed to output growth in the Czech Republic, Hungary, Poland, and Slovenia.

Let us elaborate some on the importance of the results of CA. Hoskisson et al. (2000) state that even within the same geographic region, emerging market economies are not homogenous, and the differences between the countries make comparisons in small samples problematic. The results of CA, first, confirm that our set of 18 TEs is, indeed, not homogenous. Second, the results demonstrate that the heterogeneity of the sample established in this study is very specific to the telecommunication industry. Meaning, it may be possible to produce entirely different groupings in the case of a comparison of 18 TEs in terms of the investments in and revenues from, let us say, international tourism, or research and development. It may even be that the 18 TEs are homogenous in some regard, such as, let us suppose, percentage of paved roads. However, we aimed to test the null hypothesis regarding the homogeneity of 18 TEs in terms of the investments in and revenues from the investments in telecoms, and we rejected it based on the results of CA. For the convenience of the reader, we provide summarized results in Table 25.3.

## Results: Decision Tree

In this part of our analysis, we use DT to identify the characteristics of those TEs, which are the most efficient in utilizing their investments in telecoms. First, we identified the largest (in terms of the number of variables) set of the data that *Yearbook of Statistics* and *WDI Database* could yield (complete list of the variables is provided in Section 3.3.1).

Second, we created a binary dummy variable, which was set as a "target" of the DT analysis. We assigned the values of "1" of the target variable to the countries comprising the "Leaders" cluster, and we assigned the values of "0" to the members of the "Majority" cluster.

The third step consisted of the iterative generation of multiple DT models, where every iteration involved the following three steps. The first step consisted of generating a DT model. The second step involved identifying the variable that was used for the top split, as well as recording the split value for that variable. During the third step, the top split variable of a current iteration was taken out from the data set (this was obtained by setting its status to "don't use" in the Decision Tree node of EM). Then the process was repeated.

In our evaluation of the resulting models, we were looking for those variables, splits along which resulted in the cleanest possible separation of the data set according to the value of the target variable. Consequently, after analysis of the generated DT models, we ended up with 15 variables that vividly differentiate the "Leaders" from the "Majority". Moreover, we have calculated the average value of the split variable, as well as the (approximate) percentile within which the value of the split falls.

The results of the DT analysis suggest that on average "Leaders" have higher:

- GDP per capita
- level of annual telecom investment
- level of international telecom traffic
- number of mobile phones
- number of telephone mainlines
- percentage of the internet users among the population
- number of teachers per pupil in the system of primary education
- percentage of the total labor force employed as R&D technicians
- level of spending on health care

At the same time, "Leaders" have lower than the "Majority":

- level of military expenditure
- percentage of the labor force serving in the military.

All the compiled information allows us to conclude that the "Leaders" appear to be wealthier, in general, than the "Majority", having better infrastructure and smaller armies. In addition to providing some new insights, our findings are congruent with the results of the previous research conducted in the context of TEs.

Cornia and Popov (2001) state that "socialist economies differed considerably among each other in terms of the military sector" and suggest that the resulting "structural distortions" affect the ability of TEs to sustain output during the transition. Results of our inquiry confirm that higher levels of militarization are associated with lower levels of macroeconomic output. Carment (2001) points out that the excessive military expenditures are associated with the reduction of investments in social sectors. Our findings demonstrate that higher levels of military expenditure of the "Majority" are associated with their lower levels of spending on health care. Campos and Coricelli (2002) state that a "quality of infrastructure fundamental for the functioning of a market economy, proxied by telephone lines" and human capital play an important role in the macroeconomic growth of TEs. Again, our investigation found evidence that TEs with better infrastructure and higher quality of human capital tend to have higher levels of macroeconomic output.

Based on the results of the DT analysis, we summarize a list of the contributing factors affecting the efficiency of utilization of investments in telecoms, as well as a per factor distribution of the "Majority" and the "Leaders", in Table 25.4.

We also provide the averages for each group and the levels of split in terms of the absolute values for each variable in Table 25.5. The information that was obtained from the analysis of each of the DT models presented as a graph in Figure 25.1.

The list of contributing factors affecting the efficiency of utilization of investments in telecoms yielded by DT analysis can be reduced to a smaller number of general factors. One of the possible groupings is presented in Table 25.6.

**TABLE 25.4**
**List of the Contributing Factors and Corresponding Distribution per Group of TEs**

| Variable/Contributing Factor | Distribution | |
|---|---|---|
| | Majority | Leaders |
| GDP per capita(constant 2,000 US $) | 97% are in bottom 60% | 90% are in top 40% |
| Annual Telecom Investment (% of GDP) | 62% are in bottom 43% | 98% are in top 57% |
| Fixed line and mobile phone subscribers | 94% are in bottom 75% | 63% are in top 25% |
| International telecom, outgoing traffic, minutes per subscriber(per 1,000 people) | 88% are in bottom 55% | 90% are in top 45% |
| Telephone mainlines(per 1,000 people) | 75% are in bottom 43% | 100% are in top 57% |
| Telephone mainlines per employee | 65% are in bottom 37% | 100% are in top 63% |
| Health Expenditure, Public (% of GDP) | 100% are in bottom 75% | 67% are in top 25% |
| Health Expenditure per Capita(Current US $) | 88% are in bottom 50% | 100% are in top 50% |
| Military personnel(% of labor force) | 72.5% are in top 40% | 100% are in bottom 60% |
| Military expenditure (% of GDP) | 63% are in top 36% | 100% are in bottom 64% |
| Pupil-teacher ratio(primary) | 100% are in bottom 70% | 63% are in top 30% |
| Internet users (per 1,000 people) | 75% are in bottom 45% | 93% are in top 55% |
| Mobile phones(per 1,000 people) | 60% are in bottom 34% | 100% are in top 66% |
| Technicians in R&D(% of labor force) | 50% are in bottom 27% | 100% are in top 73% |
| R&D expenditure(% of GDP) | 67% are in bottom 45% | 83% are in top 55% |

**TABLE 25.5**
**List of the Contributing Factors, Averages in Absolute Values, and Levels of Split**

| Variable/Contributing Factor | Averages (In Absolute Values) | | Level of Split |
|---|---|---|---|
| | Majority | Leaders | |
| GDP per capita(constant 2,000 US $) | 1034.49 | 4636.11 | 2720.51 |
| Annual Telecom Investment (% of GDP) | 0.60 | 1.11 | 1.11 |
| Fixed line and mobile phone subscribers (per 1,000 people) | 158.78 | 485.24 | 351.27 |
| International telecom, outgoing traffic, minutes per subscriber | 95.13 | 98.86 | 117.08 |
| Telephone mainlines(per 1,000 people) | 160.46 | 296.63 | 249.96 |
| Telephone mainlines per employee | 71.06 | 154.89 | 122.90 |
| Health Expenditure, Public (% of GDP) | 3.15 | 5.05 | 5.00 |
| Health Expenditure per Capita(Current US $) | 77.05 | 383.27 | 151.00 |
| Military personnel(% of labor force) | 1.12 | 1.95 | 1.64 |
| Military expenditure (% of GDP) | 2.27 | 1.95 | 2.09 |
| Pupil-teacher ratio(primary) | 18.57 | 14.22 | 15.71 |
| Internet users (per 1,000 people) | 114.55 | 27.72 | 22.86 |
| Mobile phones(per 1,000 people) | 329.70 | 73.37 | 81.03 |
| Technicians in R&D(% of labor force) | 0.06 | 0.11 | 0.048 |
| R&D expenditure(% of GDP) | 0.52 | 0.87 | 0.57 |

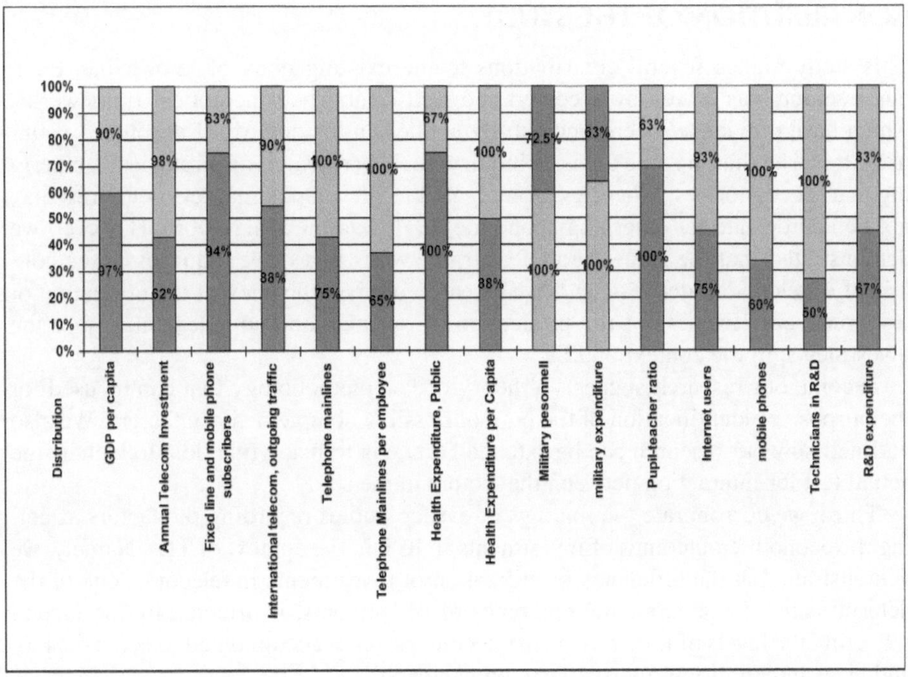


**FIGURE 25.1** "Majority" (dark grey) vs. "Leaders" (light grey): comparison in terms of 15 criteria.

**Legend: "GDP per capita" – 90% of the "Leaders" are in the top 40% while 97% of the "Majority" are in the bottom 60%.**

## TABLE 25.6
## General Factors Contributing to Efficiency of Utilization of ICT Investment in TEs

| General Factors | Contributing Factor |
|---|---|
| Level of economic development | GDP per capita (constant 2,000 US $) |
| Level of investments in telecoms | Annual telecom investment (% of GDP) |
| Level of accumulated ICT capital | Telephone mainlines (per 1,000 people) telephone mainlines per employee mobile phones (per 1000 people) |
| Level of utilization of accumulated ICT capital | Fixed line and mobile phone subscribers international telecom, outgoing traffic, minutes per subscriber internet users (per 1,000 people) |
| Level of socio-technical development | Health expenditure, public (% of GDP) health expenditure per capita (Current US $) pupil-teacher ratio (primary) technicians in R&D (% of labor force) R&D expenditure (% of GDP) |
| Level of militarization | Military personnel (% of labor force) military expenditure (% of GDP) |

## CONTRIBUTION OF THE STUDY

This study makes several contributions to the existing body of knowledge. First, our research was driven by accepted and well-established theoretical frameworks, which have previously been successfully applied in the context of developed countries. It is true that in some areas of inquiry, such as research on strategies in emerging market economies, "theories promulgated for developed market economies may not be appropriate for emerging economies" (Hoskisson et al., 2000). However, we demonstrated that the same theoretical frameworks that drive inquiries in the context of developed countries can drive research on the efficiency of the utilization of resources, effectiveness of the production of revenue, and complementarity of the investments, in the context of TEs.

Second, our research suggests a theory-driven methodology that can be used for the purpose of identification of the pool of possibly complementary factors. We also outlined how our research can be extended (i.e., via translog function) to include the actual test for interaction between the factors in the study.

Third, we corroborate the findings of earlier studies regarding the factors affecting the economic outcomes of investments in ICT in the context of TEs. Namely, we demonstrate that the efficiency of utilization of investments in telecoms, one of the determinants of the economic outcomes of investments, is affected by the factors reflecting the levels of investments in telecoms, level of accumulated telecom capital, and level and quality of the available human capital.

Fourth, in addition to corroborating the results of the previous studies, our research obtained some new empirical findings in the form of a set of factors affecting the level of the economic outcome of investments in telecoms. Namely, we were able to demonstrate that the overall level of economic development, the level of utilization of the accumulated telecom capital, and the level of militarization of the economy are among the factors that affect the efficiency of the utilization of investments in telecoms.

Fifth, from the theoretical standpoint, this research provides a contribution to the existing body of knowledge by suggesting the additional set of variables that should be included in the model describing the relationship between investments in ICT and the economic outcomes of such investments. This study also sheds some new light regarding the type of complementary investments that must take place in parallel with investments in ICT.

The results of the study could be of benefit to the community of practitioners as well. Any policy-maker or investor shall be well served by taking into consideration the multiple factors affecting the economic outcomes of investments in telecoms, acknowledging, therefore, that no significant increase in the level of investments in telecoms shall directly result a similar increase in revenues from such investments.

Moreover, the results of the study also suggest that an increase in efficiency of the utilization of the resources should go hand-in-hand with the increase in investments and vice versa. Meaning, any significant increase in investments in telecoms should be accompanied by the increase in the efficiency of utilization of these investments, while it is unlikely that any significant increase in efficiency of utilization of resources can take place without any additional investments.

This bears an important implication on strategy of investments in telecoms in the context of TEs. The highly industrialized and efficient (in terms of the utilization of resources) TEs can handle, effectively and efficiently, one-time large investments. However, the investments in the context of the less industrialized TEs, characterized by inefficient utilization of resources, should be made gradually, in step-by-step fashion. Consequently, the results of each investment step must be evaluated against the criteria of increasing efficiency of utilization of investments before any additional investments are made.

## SUMMARY AND CONCLUSION

In this study, we searched for some of the factors affecting the efficiency of utilization of investments in telecoms in the context of 18 TEs. The use of DEA allowed us to determine the relative efficiency of the utilization of investments by each TE in the sample. By using CA, we were able to demonstrate the presence of the two heterogeneous subgroups within our sample of 18 TEs. By incorporating the results of CA with the results of DEA, we determined the presence of a significant difference between the two groups of TEs in terms of the relative efficiency of the utilization of investments in telecoms. We named the subgroup with the higher averaged relative efficiency the "Leaders" and the group with the lower averaged relative efficiency the "Majority".

By using DT, we were able to demonstrate that the "Leaders" differs from the "Majority" in terms of 15 factors that reflect the differences in the level of economic development, level of investments in telecoms, level of accumulated telecom capital, as well as the level of utilization of telecom capital, level of socio-technical development, and level of militarization of TE. The results of our study strongly suggest that these factors affect the level of the efficiency of the utilization of investments in telecoms in the context of TEs.

Based on the results of our study, we hypothesize that the "majority" has a lower level of relative efficiency of utilization of investments in telecoms because of the three following reasons: first, the "majority", in comparison with the "leaders", simply does not invest enough in telecoms to be concerned with the issue of efficiency. Second, even if the "majority" does invest at a relatively sufficient level, a large part of the investments is directed not toward obtaining revenue, but toward building the required supporting infrastructure. Because only a part of the overall investments is involved in revenue production, it is reflected as an overall low-efficiency score. Third, as the level of the telecom infrastructure increases, it brings about an increased complexity associated with the utilization of the infrastructure. Consequently, the process by which investments are transformed into revenues becomes more complicated, and the "majority" lacks the necessary socio-technical "know-how" to manage that increased complexity.

Testing of these hypotheses represents the outline of some of the possible directions for future research in this area. Intuitively, it would make sense to expect a gradual increase in investments as being accompanied by an associated gradual increase in the level of learning regarding the utilization of the investments, in the form of the socio-technical "know-how". However, at this point it is not clear whether the increase in efficiency brings about the increase in investments, or whether the additional investments allow for obtaining the higher efficiency of the utilization of the investments.

Despite the contributions that this study makes, our research is not without its limitations. First, the use of the different clustering criteria may produce a clustering solution that is different from the one obtained in this research. Second, during DEA we used a model where all the input variables and the output variables were weighted equally. However, some of the variables may be more important than other variables and, therefore, such variables perhaps should be weighted more heavily than other variables. Moreover, while we provided a rationalization for the DEA model used in our study, ultimately, there are no objective criteria according to which such model can be constructed. Hence, a different DEA model could result in different scores of relative efficiency.

Another limitation of this study is associated with the quantity of the data. Clearly, our research may have offered richer insights if more variables were available from the data sources that we used. This limitation, however, is not unique to our study but is characteristic to the research in this area in general (Hoskisson et al. 2000). In addition, it may be argued that the time series data covering a 10-year period could be insufficient to inquire into the nature of the events taking place on a macro-economic level. Nevertheless, in the area of research where it "would appear to be a major need for longitudinal studies" (Hoskisson et al., 2000), we feel that the contributions provided by our study outweigh its limitations.

## ACKNOWLEDGMENT

Material in this chapter previously appeared in: Contributing Factors to Information Technology Investment Utilization in Transition Economies: An Empirical Investigation, *Information Technology for Development, 14(1),* 52–75.

## REFERENCES

Aldenderfer, M.S., & Blashfield, R.K. (1984). Cluster Analysis. Sage, Newbury Park, CA.

Ali, A.I. (1994). Computational aspects of DEA. In: Charnes, A., Cooper, W.W., Lewin, A. & Seiford, L.M. (eds.). Data Envelopment Analysis: Theory, Methodology and Applications. Kluwer Academic Publishers, Boston, MA, pp. 63–88.

Athey, S., & Schmutzler, A. (1995). Product and process flexibility in an innovative environment. RAND Journal of Economics, 26(4), 557–574.

Breiman, L., Friedman, J.H., Olshen, R.A., & Stone, C.J. (1984). Classification and Regression Trees. Chapman & Hall, New York, NY.

Bugamelli, M., & Pagano, P. (2004). Barriers to iin ICT. Applied Economics, Taylor and Francis Journals, 36(20), 2275–2286.

Campos, N., & Coricelli, F. (2002). Growth in transition: What we know, what we don't and what we should. Journal of Economic Literature, American Economic Association, 40(3), 793–836.

Charnes, A., Cooper, W. W., Lewin, A. Y., & Seiford, L. M. (Eds.). (1994). Data Envelopment Analysis: Theory, Methodology, and Applications. Kluwer, Boston, MA.

Colecchia, A., & Schreyer, P. (2001). The Impact of Information Communications Technology on Output Growth. STI Working Paper 2001/7. OECD, Paris.

Cooper, W.W., Seiford, L.M., & Zhu, Joe. (2004). Data envelopment analysis: history, models and interpretations. In: W.W. Cooper, L.M. Seiford and J. Zhu. (Eds.). Handbook on Data Envelopment Analysis. Kluwer Academic Publishers, Boston, MA, pp. 1–39.

Cornia, G.A., & Popov, V. (2001). Structural and institutional factors in the transition to the market economy: An overview. In: Cornia, G.A. & Popov, V. (Eds.). Transition and Institutions: The Experience of Late Reformers. Oxford University Press, New York, NY, pp. 3–29.

Daveri, F. (2000). Is Growth an Information Technology Story in Europe too? Working Paper. University of Parma, Parma, Italy.

Daveri, F. (2002). The new economy in Europe, 1992–2001. Oxford Review of Economic Policy, 18(3), 345–362.

Dewan, S., & Kraemer, K. (1998). International dimensions of the productivity paradox. Communications of the ACM, 41(8), 56–62.

Dewan, S., & Kraemer, K. (2000). Information technology and productivity: evidence from country level data. Management Science (special issue on the Information Industries), 46(4), 548–562.

Edgeworth, F.Y. (1881). Mathematical Psychics. Kegan Paul, London.

Gera, S., & Wulong, Gu. (2004). The effect of organizational innovation and information and communications technology on firm performance. International Productivity Monitor, Centre for the Study of Living Standards, 9, 37–51.

Giuri, P., Torrisi S., & Zinovyeva, N. (2005). ICT, Skills and Organizational Change: Evidence from a Panel of Italian Manufacturing Firms. LEM Papers Series 2005/11. Laboratory of Economics and Management (LEM), Sant'Anna School of Advanced Studies, Pisa, Italy.

Hernando, I., & Nunez, S. (2002). The Contribution of ICT to Economic Activity: A Growth Accounting Exercise with Spanish Firm-Level Data. Banco de España – Servicio de Estudios Documento de Trabajo No. 0203. Available at: http://www.bde.es/informes/be/docs/dt0203e.pdf.

Hoskisson, R., Eden, L., Lau, C., & Wright, M. (2000). Strategy in emerging economies. Academy of Management Journal, 43(3), 249–267.

Jalava, J., & Pohjola, M. (2002). Economic growth in the new economy: Evidence from advanced economies. Information Economics and Policy, 14(2), 189–210.

Jorgenson, D.W. (2001). Information technology and the US economy. American Economic Review, 91(March), 1–32.

Jorgenson, D.W., & Stiroh, K.J. (2000). Raising the speed limit: US economic growth in the information age. Brookings Papers on Economic Activity, 2(1), 125–211.

Lambertini, L. (2003). The monopolist's optimal R&D portfolio. Oxford Economic Papers, 55, 561–78.

Lin, P., & Saggi, K. (2002). Product differentiation, process R&D, and the nature of market competition. European Economic Review, 46, 201–211.

Lins, M.P.E., Gomes, E.G., de Mello, J.C.C.B.S., & de Mello, A.J.R.S. (2003). Olympic ranking based on a zero sum gains DEA model. European Journal of Operational Research, 148(2), 312–322.

Loukis, E., & Sapounas, I. (2004). The Impact of Information Systems Investment and Management on Business Performance in Greece. ECIS 2005 Conference paper.

Murakami, T. (1997). The Impact of ICT on Economic Growth and the Productivity Paradox. Available on-line at: http://www.tcf.or.jp/data/19971011_Takeshi_Murakami_2.pdf.

Myers, D. (2004). Construction Economics. Spon Press, UK.

OECD (2004). DAC Network on Poverty Reduction: ICTs and Economic Growth in Developing Countries. OECD, Paris. Available on-line at: http://www.oecd.org/dataoecd/15/54/34663175.pdf.

OECD (2005). Good practice paper on ICTs for economic growth and poverty reduction. The DAC Journal, 6(3).

Oliner, S.D., & Sichel, D.E. (2000). The resurgence of growth in the late 1990s: Is information technology story? Journal of Economic Perspectives, 14(4), 3–22.

Oliner, S.D., & Sichel, D.E. (2002). Information technology and productivity: Where are we now and where are we going? Economic Review, 3(3), 15–41.

Ollman, B. (1997). Market Socialism: The Debate Among Socialists. Routledge, UK.

Piatkowski, M. (2002). The 'New Economy' and Economic Growth in Transition Economies. WIDER Discussion Paper No. 2002/63. WIDER, Helsinki.

Piatkowski, M. (2003). Does ICT Investment Matter for Output Growth and Labor Productivity in Transition Economies? TIGER Working Paper Series, No. 47. Available on-line at: www.tiger.edu.pl.

Rosenkranz, S. (2003). Simultaneous choice of process and product innovation when consumers have a preference for product variety. Journal of Economic Behavior & Organization, 50(2), 183–201.

Schreyer, P. (2000). The Contribution of Information and Communication Technology to Output Growth: a Study of the G7 Countries. STI Working Papers 2000/2. OECD, Paris.

Solow, R. (1957). Technical change and the aggregate production function. Review of Economics and Statistics, 39(3), 312–20.

Smith, A. (1776). The Wealth of Nations. Available on-line at: http://www.marxists.org/reference/archive/smith-adam/works/wealth-of-nations/.

Stiroh, K. (2002). Information technology and the U.S. productivity revival: What do the industry data say? American Economic Review, 92(5), 1559–1576.

United Nations Conference and Development [UNCTAD]. (2004). The Least Developed Countries Report. United Nations Publications.

United Nations Development Program [UNDP]. (1998). Human Development Report. Oxford University Press, New York, NY.

UN ICT Task Force Report. (2005). Innovation and Investment: Information and Communication Technologies and the Millennium Development Goals. Report Prepared for the United Nations ICT Task Force in Support of the Science, Technologies and Innovation Task Force of the United Nations Millennium Project. Available on-line at: www.unicttaskforce.org/.

United Nations Statistics Division [UNSD]. (1999). Standard Country or Area Codes for Statistical Use. Statistical Papers, Series M, No.49 Rev. 4. United Nations, New York, NY.

Van Ark, B., Melka, J., Mulder, N., Timmer, M., & Ypma, G. (2002). ICT Investments and Growth Accounts for the European Union, 1980–2000. Research Memorandum GD-56, Groningen Growth and Development Centre, Groningen. Available on-line at: www.eco.rug.nl/ggdc/homeggdc.html.

World Bank. (2002). Global Economic Prospects and the Developing Countries. World Bank. Available on-line at: http://web.worldbank.org/WBSITE/EXTERNAL/EXTDEC/EXTDECPROSPECTS/GEPEXT/EXTGEP2002.

WT/ICT Development Report. (2006). Measuring ICT for Social and Economic Development. International Telecommunication Union's World Telecommunication/ICT Development Report, 8th edition. Available on-line at: http://www.itu.int/ITU-D/ict/publications/wtdr_06/index.html.

Yearbook of Statistics. (2004). Telecommunication Services Chronological Time Series 1993–2002. ITU Telecommunication Development Bureau (BDT), International Telecommunication Union. Available on-line at: www.itu.int/ITU-D/ict/publications.

## APPENDIX A

### TABLE A.1
### Farrel Input-Saving Measure of Efficiency, CRS

| Country | 1993 | 1994 | 1995 | 1996 | 1997 | 1998 | 1999 | 2000 | 2001 | 2002 |
|---|---|---|---|---|---|---|---|---|---|---|
| **Albania** | 1 | 1 | 1 | 1 | 1 | 1 | 1 | 0.80 | 1 | 1 |
| **Armenia** | 0.79 | 1 | 1 | 1 | 1 | 0.89 | 0.81 | 0.73 | 0.86 | 0.80 |
| **Azerbaijan** | 0.49 | 0.60 | 1 | 1 | 1 | 0.88 | 0.51 | 0.84 | 0.49 | 0.48 |
| **Belarus** | 0.56 | 0.31 | 0.68 | 0.64 | 0.56 | 0.47 | 0.57 | 0.44 | 0.39 | 0.52 |
| **Bulgaria** | 1 | 1 | 1 | 0.88 | 0.86 | 0.83 | 0.92 | 1 | 0.72 | 1 |
| **Czech Republic** | 1 | 1 | 0.78 | 0.60 | 0.81 | 0.97 | 0.98 | 0.81 | 0.68 | 0.90 |
| **Estonia** | 0.81 | 0.79 | 1 | 1 | 1 | 1 | 1 | 1 | 1 | 1 |
| **Hungary** | 1 | 1 | 1 | 1 | 1 | 1 | 1 | 1 | 1 | 1 |
| **Kazakhstan** | 0.80 | 0.40 | 0.60 | 0.67 | 0.75 | 0.65 | 0.51 | 0.54 | 0.51 | 1 |
| **Kyrgyzstan** | 1 | 1 | 1 | 0.72 | 0.56 | 1 | 1 | 1 | 1 | 1 |
| **Latvia** | 0.41 | 0.99 | 0.94 | 0.94 | 1 | 0.85 | 0.69 | 0.67 | 0.55 | 0.52 |
| **Lithuania** | 0.82 | 0.82 | 0.74 | 0.85 | 0.45 | 0.46 | 0.50 | 0.46 | 0.55 | 0.96 |
| **Moldova** | 0.22 | 1 | 1 | 1 | 1 | 1 | 1 | 1 | 1 | 1 |
| **Poland** | 0.93 | 1 | 0.90 | 0.76 | 0.85 | 1 | 0.84 | 1 | 1 | 1 |
| **Romania** | 0.55 | 0.52 | 0.62 | 0.59 | 0.48 | 0.46 | 0.55 | 0.52 | 0.63 | 0.90 |
| **Slovakia** | 0.82 | 0.80 | 0.79 | 0.87 | 0.55 | 0.51 | 0.71 | 1 | 1 | 1 |
| **Slovenia** | 1 | 1 | 1 | 1 | 1 | 1 | 1 | 0.71 | 0.81 | 1 |
| **Ukraine** | 1 | 0.70 | 0.70 | 1 | 0.92 | 1 | 1 | 1 | 0.85 | 0.88 |

### TABLE A.2
### Farrel Input-Saving Measure of Efficiency, VRS

| Country | 1993 | 1994 | 1995 | 1996 | 1997 | 1998 | 1999 | 2000 | 2001 | 2002 |
|---|---|---|---|---|---|---|---|---|---|---|
| **Albania** | 1 | 1 | 1 | 1 | 1 | 1 | 1 | 0.95 | 1 | 1 |
| **Armenia** | 1 | 1 | 1 | 1 | 1 | 0.89 | 0.83 | 0.75 | 0.87 | 0.80 |
| **Azerbaijan** | 0.82 | 0.92 | 1 | 1 | 1 | 0.95 | 1 | 1 | 1 | 1 |
| **Belarus** | 0.75 | 0.59 | 0.77 | 0.72 | 0.64 | 0.60 | 0.64 | 0.67 | 0.66 | 0.77 |
| **Bulgaria** | 1 | 1 | 1 | 0.93 | 0.95 | 0.83 | 1 | 1 | 0.72 | 1 |
| **Czech Republic** | 1 | 1 | 0.80 | 0.75 | 0.87 | 0.97 | 1 | 0.95 | 0.74 | 0.94 |
| **Estonia** | 0.86 | 0.83 | 1 | 1 | 1 | 1 | 1 | 1 | 1 | 1 |
| **Hungary** | 1 | 1 | 1 | 1 | 1 | 1 | 1 | 1 | 1 | 1 |
| **Kazakhstan** | 1 | 0.48 | 0.62 | 0.70 | 1 | 0.65 | 0.70 | 0.78 | 0.75 | 1 |
| **Kyrgyzstan** | 1 | 1 | 1 | 1 | 1 | 1 | 1 | 1 | 1 | 1 |
| **Latvia** | 0.57 | 1 | 1 | 0.95 | 1 | 0.93 | 0.90 | 0.94 | 0.83 | 0.76 |
| **Lithuania** | 0.87 | 0.93 | 0.82 | 0.87 | 0.60 | 0.63 | 0.81 | 0.91 | 0.86 | 1 |
| **Moldova** | 1 | 1 | 1 | 1 | 1 | 1 | 1 | 1 | 1 | 1 |
| **Poland** | 0.93 | 1 | 0.91 | 0.85 | 0.93 | 1 | 0.96 | 1 | 1 | 1 |
| **Romania** | 0.61 | 0.63 | 0.67 | 0.66 | 0.63 | 0.65 | 0.73 | 0.75 | 0.72 | 0.91 |
| **Slovakia** | 0.84 | 0.82 | 0.80 | 0.87 | 0.68 | 0.66 | 0.74 | 1 | 1 | 1 |
| **Slovenia** | 1 | 1 | 1 | 1 | 1 | 1 | 1 | 0.96 | 0.87 | 1 |
| **Ukraine** | 1 | 0.70 | 0.75 | 1 | 1 | 1 | 1 | 1 | 0.86 | 1 |

## TABLE A.3
## Farrel Input-Saving Measure of Efficiency, NIRS

| Country | 1993 | 1994 | 1995 | 1996 | 1997 | 1998 | 1999 | 2000 | 2001 | 2002 |
|---|---|---|---|---|---|---|---|---|---|---|
| **Albania** | 1 | 1 | 1 | 1 | 1 | 1 | 1 | 0.80 | 1 | 1 |
| **Armenia** | 0.79 | 1 | 1 | 1 | 1 | 0.89 | 0.81 | 0.73 | 0.86 | 0.80 |
| **Azerbaijan** | 0.49 | 0.60 | 1 | 1 | 1 | 0.88 | 0.51 | 0.84 | 0.49 | 0.48 |
| **Belarus** | 0.56 | 0.31 | 0.68 | 0.64 | 0.56 | 0.47 | 0.57 | 0.44 | 0.39 | 0.52 |
| **Bulgaria** | 1 | 1 | 1 | 0.93 | 0.95 | 0.83 | 1 | 1 | 0.72 | 1 |
| **Czech Republic** | 1 | 1 | 0.78 | 0.60 | 0.81 | 0.97 | 1 | 0.81 | 0.68 | 0.90 |
| **Estonia** | 0.81 | 0.79 | 1 | 1 | 1 | 1 | 1 | 1 | 1 | 1 |
| **Hungary** | 1 | 1 | 1 | 1 | 1 | 1 | 1 | 1 | 1 | 1 |
| **Kazakhstan** | 0.80 | 0.40 | 0.60 | 0.67 | 1 | 0.65 | 0.51 | 0.54 | 0.51 | 1 |
| **Kyrgyzstan** | 1 | 1 | 1 | 0.72 | 0.56 | 1 | 1 | 1 | 1 | 1 |
| **Latvia** | 0.41 | 1 | 1 | 0.94 | 1 | 0.85 | 0.69 | 0.67 | 0.55 | 0.52 |
| **Lithuania** | 0.82 | 0.93 | 0.74 | 0.87 | 0.45 | 0.46 | 0.50 | 0.46 | 0.55 | 0.96 |
| **Moldova** | 0.22 | 1 | 1 | 1 | 1 | 1 | 1 | 1 | 1 | 1 |
| **Poland** | 0.93 | 1 | 0.90 | 0.76 | 0.85 | 1 | 0.84 | 1 | 1 | 1 |
| **Romania** | 0.55 | 0.52 | 0.62 | 0.59 | 0.48 | 0.46 | 0.55 | 0.52 | 0.63 | 0.90 |
| **Slovakia** | 0.82 | 0.80 | 0.79 | 0.87 | 0.55 | 0.51 | 0.74 | 1 | 1 | 1 |
| **Slovenia** | 1 | 1 | 1 | 1 | 1 | 1 | 1 | 0.71 | 0.81 | 1 |
| **Ukraine** | 1 | 0.70 | 0.70 | 1 | 1 | 1 | 1 | 1 | 0.85 | 1 |

# 26 Increasing the Discriminatory Power of DEA in the Presence of the Sample Heterogeneity with Cluster Analysis and Decision Trees

## INTRODUCTION

Data envelopment analysis (DEA) is a non-parametric data analytic technique that is extensively used by various research communities (e.g., Hong et al., 1999; Sohn and Moon, 2004; Seol et al., 2007) since its introduction by Charnes et al. (1978). The domain of inquiry of DEA is a set of the entities, commonly called decision making units (DMUs), which receive multiple inputs and produce multiple outputs. Given a sample of the DMUs, the purpose of DEA is establishing of the relative efficiencies of each DMU within a sample. By collapsing the multiple inputs and outputs into the meta input and meta output, DEA employs linear programming techniques to establish an input–output based ratio that is then used to derive and assign a relative efficiency score to each DMU in the sample.

As any non-parametric method, DEA eschews the hard-to-satisfy distributional assumption of data normality, thus allowing to investigator to inquire into the nature of the relative efficiency of the sample of convenience. This comes with the price, however, for DEA relies on a different set of assumptions that must be satisfied for the results to be valid.

One of the fundamental assumptions of DEA is that of a functional similarity of the DMUs in a sample. This simply means that in order to compare, meaningfully, the relative efficiencies of DMUs in the sample (or data set), these DMUs must be similar in terms of utilization of the inputs and production of the outputs. Thus, DEA requires us to make sure that we compare apples and apples, and not apples and oranges.

The difficulty arises when DMUs are represented not by apples and oranges, but by somewhat less homogenous entities such as people, companies, schools, and countries.

In general, departments are more complex than people, companies are more complex than departments, and industries and countries are more complex than companies. As a complexity of a DMU increases, it gives rise to the increase in diversity in the set of DMUs, and this leads, problematically for DEA, to heterogeneity, rather than homogeneity, of the sample.

One of our motivations for addressing is this issue is that it arises in various applications of the DEA methodology to real world problems. In our case we were exploring the relative efficiencies of a set of countries that were transitioning from centralized, planned economies to the market economies. To the extent that they are often considered by the international community, including significant actors such as the World Bank and IMF, to be a single group under the label *Transition Economies* (TE), there appears to be good reason to meaningfully compare them using DEA. However, in the situations where a domain of DMUs has been superimposed by the existing classifications, i.e., TEs, firms, departments, industries, one may end up with the sample consisting of DMUs that are functionally similar, yet heterogeneous. The DMUs are functionally similar because they receive, according to the specified DEA model, the same set of the inputs and produce the same set of the outputs, and such DMUs are possibly heterogeneous because no DMU could be specified completely by its inputs and the outputs. Again, complexity of the DMUs gives rise to the heterogeneity of the relevant set.

It would appear that while the functional similarity of DMUs in a sample is assumed, the homogeneity of the DMUs is taken for granted. This, of course, severely limits the discriminatory power of DEA results. On the other hand that the explicit consideration of the possible heterogeneity in a sample would increase the discriminatory power of the results of the DEA.

Currently, it would appear that the test of the assumption of the homogeneity of the DMUs in the sample represents the less rigorous, if not non-existent, part of DEA, for it is implicitly assumed that an investigator made sure that this important assumption holds. We suggest that the rigor of DEA, as well as its discriminatory power, could be increased by making the process of the assumption checking more explicit and objective.

If one intends to inquire into the differences between efficient and inefficient DMUs in the sample, then, fundamentally, there are only two possible routes. First, if the assumption of the homogeneity holds, then the differences in the scores of the relative efficiency could be investigated without any adjustments. Or, second, if the assumption of the homogeneity of the DMUs in the sample does not hold, then the differences in the scores of the relative efficiency should be investigated with the appropriate adjustments. The question, then, becomes: *What are the appropriate adjustments?*

It is not our intent to propose that DEA should only be performed in the situations where homogeneity of the DMUs has been decisively established. After all, a homogeneity, or, heterogeneity, of the DMUs in a sample is a matter of a degree. Sometimes the differences are minor and the researchers should declare that essentially the assumption of the homogeneity of the DMUs holds. Other times, however, the researchers might want to take into consideration the differences between the DMUs, while still comparing their relative efficiencies within the same sample.

Currently, to our knowledge, there exists no methodology allowing a researcher to investigate the differences between the relatively efficient and inefficient DMUs while taking into consideration heterogeneity of the DMUs in the sample. In this research we aim to address this problem. Namely, we propose and illustrate a three-step methodology allowing a researcher, first, to inquire into the differences between the DMUs in the sample, second, conduct DEA, and third, inquire into the differences between the relatively efficient and inefficient DMUs while taking into consideration the differences uncovered in the first step.

The first step of our methodology utilizes cluster analysis (CA), the second step employs DEA, and the third step relies on the decision tree (DT) analysis. We provide brief overview of these data analytic techniques next, followed by the description of the proposed methodology.

## THE PROPOSED METHODOLOGY

In this section, we describe, in step-by-step fashion, the sequence of the procedures constituting the proposed methodology. Before describing our methodology, we provide an overview of the data set that is used in our running illustrative example.

### Overview of Data Set of Illustrative Example

To illustrate our methodology in action, we have chosen the following problem.

Given a 10-year data set on 18 TEs, spanning a period from 1993 to 2002, we want to find out what are:

1. The differences in the relative efficiencies of these economies regarding their investments in telecoms, and
2. Some of the factors that contribute to the differences in the relative efficiencies.

The data for this study were obtained from two sources. The first source was represented by the database of World Development Indicators, which is the World Bank's comprehensive database on development data. The second source of the data was represented by the *Yearbook of Statistics,* which is published yearly by International Telecommunication Union (ITU). In our choice of variables, we were greatly restricted by the availability of the data. For example, while the development data of the World Bank's database covers more than 600 indicators for 208 economies, data on many of the indicators relevant to our research were not available, or were available only for a few countries, or contained too few data points to be useful in statistical analysis. In terms of the length of the time series, we were restricted to the period from 1992 to 2002, data for which were provided by *Yearbook of Statistics* of ITU.

In our choice of TEs we were guided by the intent to isolate a group of countries that started the process of transition in approximately the same time. As a result, we have decided to concentrate on the 25 countries of the former Soviet bloc.

**TABLE 26.1**
**List of Variables for DEA Models**

| Role | Sub-set of Variables |
|---|---|
| Input | GDP per capita (in current US $),<br>Full-time telecommunication staff (% of total labor force),<br>Annual telecom investment per telecom worker,<br>Annual telecom investment (% of GDP in current US $),<br>Annual telecom investment per capita,<br>Annual telecom investment per worker |
| Output | Total telecom services revenue per telecom worker,<br>Total telecom services revenue (% of GDP in current US $),<br>Total telecom services revenue per worker,<br>Total telecom services revenue per capita |

Based on the availability of the data, the following 18 transitional economies out of 25 have been selected for this research: Albania, Armenia, Azerbaijan, Belarus, Bulgaria, Czech Republic, Estonia, Hungary, Kazakhstan, Kyrgyz Republic, Latvia, Lithuania, Moldova, Poland, Romania, Slovak Republic, Slovenia, and Ukraine. Despite the original intent, the data offered for 7 out of 25 TEs, namely, Tajikistan, Turkmenistan, Uzbekistan, Georgia, Macedonia, Russian Federation and Croatia, turned out to be insufficient to allow the inclusion of these economies in this study. For the DEA part of the methodology we have identified a model consisting of the input variables and output variables that are listed in Table 26.1.

## Description of the Methodology

Our methodology has the following three major steps that will be described in detail:

1. Determine the structural homogeneity status of the data set
2. Determine the relative efficiency status of each DMU
3. Describe the relative efficiency categories

These steps require the initialization of relevant parameters including:

- $k_{Max}$: the maximum possible number of clusters that would be of interest to the decision-maker. This parameter is required in Step 1.
- $\tau_{Outlier}$: the threshold that is used to determine if a cluster is an outlier. This parameter is required in Step 1.
- *DMU_Goal*: This could be “Input Orientation” or “Output Orientation”. This parameter is required in Step 2.
- *DMU_Criterion*: This could be CRS, VRS, or NIRS. This parameter is required in Step 2.

## Step 1: Determine the Structural Homogeneity Status of the Data Set

*Description of Step 1*

**Sub-Step 1a:**

a. Apply two-step approach to generate segmentations of sizes $k_{Max}$ through $k_{Min}$.
b. Set $k = k_{\text{Max}}$.

**Sub-Step 1b***:*

Examination the segmentation with $k$ clusters.

IF $k > 1$ and there is at least one cluster is that consists of less than $\tau_{Outlier}$ percent of the DMUs

THEN

Set $k = k - 1$;
Repeat Substep 1b

ELSE

Current Segmentation with k clusters provides the "natural" groupings of the DMUs;
Terminate Step 1.

*Justification of Step 1*

The intended purpose of this CA Step 1 of our methodology is to investigate a "structural similarity" of the data set. Structural similarity of the DMUs reflects not the types, not the transformation of the inputs into outputs, but the levels of the inputs and outputs that DMUs receive and produce. Consequently, in the first step of our methodology we aim to determine whether all DMUs in the sample are similar in terms of the levels of the received inputs and the levels of the produced outputs. It could be suggested that the purpose of CA is to test the assumption of the homogeneity of the domain given the chosen DEA model. Consequently, homogeneity of the domain of the DMUs is always going to be relative to the given DEA model.

Thus, the required pre-requisite to the first step in our methodology is that an investigator has identified a DEA model, i.e., a set of the inputs and a set of the outputs, which are going to be used in the Step 2. Once the model is determined, the actual data set that is going to be used to perform DEA is subjected to CA. During the first stage of CA we suggest to start, assuming that an investigator uses a software package allowing to conduct CA, with generating automatically a baseline clustering solution. Once it is done and the certain large number of the clusters has been generated, an investigator should evaluate the membership of each cluster. We suggest, as a rule of thumb, not to retain any cluster with the number of DMUs less than $\tau_{Outlier}$ percent of the total data set.

By gradually decreasing the number of clusters over the iterations of CA, an investigator would get one of the two types of the CA solutions. First type is reflected by the presence of a single large cluster, containing, as a rule of thumb, $(100 - \tau_{Outlier})$ percent or more of the DMUs, and one or two small clusters, with combined membership of (again as a rule of thumb) $\tau_{Outlier}$ percent or less of the sample. If this is the case, we suggest that an investigator should treat the sample of DMUs as homogenous.

In the case of the second type of CA solution, however, an investigator ends up with two or more clusters with the membership of more than 10% of the sample in each cluster. A possible case in this scenario is when the final solution is presented in the form of two or three large clusters and a single small cluster (less than 10% of the sample). In this situation we suggest a visual inspection of the solution (the diagrams are provided by most of the software packages), or examination of the distances between the clusters; based on the results of the evaluation a small cluster should be combined with the closest large cluster.

If the CA yields a second type of a solution, we suggest that the sample should be treated as heterogeneous and this heterogeneity should be reflected in labeling the resulting clusters as "Cluster1", "Cluster2", etc. Consequently, an investigator should document the membership of each cluster, by noting which DMUs belong to which cluster. After it is done, we proceed to the next step, DEA.

*Illustration of Step 1*

For our illustration of this step we decided to use following eight variables to inquire into the homogeneity of our data set:

1. Total telecom services revenue (% of GDP in current US $),
2. Total telecom services revenue per capita (Current US $),
3. Total telecom services revenue per worker (Current US $),
4. Total telecom services revenue per telecom worker (Current US $),
5. Annual telecom investment per capita (Current US $),
6. Annual telecom investment (% of GDP in current US $),
7. Annual telecom investment per worker (Current US $),
8. Annual telecom investment per telecom worker (Current US $).

The parameter $\tau_{Outlier}$ was set to 10%, $k_{\text{Max}}$ was set to 5 and $k_{\text{Min}}$ to 2. We used SAS Enterprise Miner (EM) to perform CA of the data set. The variables that we used are not measured on the same scale, so, prior to CA we transformed the data by standardizing the variables. We started our analysis by choosing "Automatic" setting, which did not require any input regarding the desired number of clusters from the researcher. Summary information regarding each obtained solution is compiled in Table 26.2.

**TABLE 26.2**
**Summary Output of Clustering**

| Number of Clusters | Number of DMUs in Each Cluster | Top-level Split Variables |
|---|---|---|
| 5 clusters | 10, 11, 20, 3, 136 | Total telecom service revenue per worker |
| | | Annual telecom investment per worker |
| | | Annual telecom investment (% of GDP) |
| 4 clusters | 10, 32, 3, 135 | Total telecom service revenue per worker |
| | | Annual telecom investment per worker |
| 3 clusters | 30, 3, 147 | Total telecom services revenue per worker |
| 2 clusters | 72, 108 | Annual telecom investment per telecom worker |

Clusters with less than $\tau_{Outlier}$ percent of the DMUs are underlined.

**TABLE 26.3**
**Cluster Membership when Number of Cluster = 2**

| Contents of the 1st Cluster | Contents of the 2nd Cluster |
|---|---|
| Albania (1993–2002) | Czech rep (1993–2002) |
| Armenia (1993–2002) | Estonia (1994–2002) |
| Azerbaijan (1993–2002) | Hungary (1993–2002) |
| Belarus (1993–2002) | Bulgaria (2002) |
| Bulgaria (1993–2001) | Latvia (1994, 1995, 1997–2002) |
| Slovak Rep (1993, 1994, 1999) | Lithuania (1999–2002) |
| Kazakhstan (1993–2002) | Slovenia (1993–2002) |
| Kyrgyz Rep (1993–2002) | Poland (1993–2002) |
| Latvia (1993, 1996) | Slovak Rep (1995–1998, 2000–2002) |
| Lithuania (1993–1998) | |
| Moldova (1993–2002) | |
| Romania (1993–2002) | |
| Ukraine (1993–2001) | |

By using CA, we were able to come up with a solution that partitions our data set into two clusters. The membership of each cluster is provided in Table 26.3. Based on the compiled information we can see, that while some of the TEs are "permanent residents" of one cluster, other TEs are "migrants", i.e., they change the cluster membership depending on a year.

Finally, we should ask ourselves the following question: What is the significance of the separation of 18 transitional economies into the two clusters? One of the possible answers is provided in the research by Piatkowski (2003), who concluded that in the period "between 1995 and 2000 ICT capital has most potently contributed to output growth in the Czech Republic, Hungary, Poland, and Slovenia". Thus, it could be suggested that we were able to separate 18 transitional economies into the two groups, one group of transitional economies that benefits the most from the investments in telecom, and another group where the benefits are less pronounced. Consequently, we labeled the members of the Cluster 1 as "*Majority*" and the members of the Cluster 2 as "*Leaders*".

## Step 2: Determine the Relative Efficiency Status of DMUs

*Description of Step 2*

Based on the values of the parameters *DMU_Goal* (i.e., "Input Orientation" or "Output Orientation") and *DMU_Criterion* (i.e., CRS, VRS, or NIRS), the relevant DEA approach is performed in order to obtain the relative efficiency scores of the DMUs in the sample. We assume that at the end of the DEA process each object is assigned a specific relative efficiency status (relatively efficient, or relatively inefficient), and so each object has an additional relative efficiency status attribute (say *EfficiencyStatus* with the value "1" for Relatively Efficient, and the value "0" for Relatively Inefficient).

*Illustration of Step 2*

For our illustration of this step, given the values of the parameters *DMU_Goal* and *DMU_Criterion*, the relevant DEA models need to be constructed.

Although not required, we generated the relative efficiencies of DMUs using both input-oriented and output-oriented DEA models, for three types of conditions, constant (CRS), variable (VRS), and non-increasing return to scale (NIRS).

Thus, we generated altogether six DEA models, with the following settings of the parameters *DMU_Goal* and *DMU_Criterion*:

1. *DMU_Goal* = "Input Orientation" and *DMU_Criterion* = CRS
2. *DMU_Goal* = "Input Orientation" and *DMU_Criterion* = VRS
3. *DMU_Goal* = "Input Orientation" and *DMU_Criterion* = NIRS
4. *DMU_Goal* = "Output Orientation" and *DMU_Criterion* = CRS
5. *DMU_Goal* = "Output Orientation" and *DMU_Criterion* = VRS
6. *DMU_Goal* = "Output Orientation" and *DMU_Criterion* = NIRS

This step does not differ in any way or form from the regular DEA, thus, no adjustments are necessary. We also computed the average relative efficiencies of the two clusters identified in the Step 1, and compare the averaged relative efficiencies of these two groups of TEs produced by the DEA. Table 26.4 demonstrates the differences in the relative efficiency between the "Leaders" and the "Majority" in terms of the utilization of the inputs. The input-orientation does not concern itself with the maximization of the outputs, but rather with maximization of the utilization of the inputs. Thus, it is probably reflective of the perspective of the policy maker, especially in the case when the available resources are limited. Table 26.5 demonstrates the differences in the relative efficiency between the "Leaders" and the "Majority" in terms of the maximization of the outputs. Unlike in the case of the input-oriented model, the output-orientation does not concern itself with the efficient utilization of the inputs, but rather with the maximization of the outputs. Thus, it is probably reflective of the perspective of the investor, especially in the case when the resources are abundant and the primary goal is to obtain the maximum revenue.

**TABLE 26.4**
**DEA: Comparison of the Clusters based on the Input-Oriented Model**

| Criterion for Comparison | "Leaders" Cluster | "Majority" Cluster | Difference | Difference % |
|---|---|---|---|---|
| Average efficiency score, CRS | 0.89 | 0.79 | 0.10 | 12.54% |
| Average efficiency score, VRS | 0.95 | 0.88 | 0.07 | 7.48% |
| Average efficiency score, NIRS | 0.89 | 0.80 | 0.09 | 11.63% |
| Average efficiency score, SE | 0.94 | 0.89 | 0.04 | 4.96% |

**TABLE 26.5**
**Comparison of the Clusters based on the Output-Oriented DEA Model**

| Criterion for Comparison | "Leaders" Cluster | "Majority" Cluster | Difference | Difference % |
|---|---|---|---|---|
| Average efficiency score, CRS | **1.17** | **1.41** | **−0.24** | **−16.71%** |
| Average efficiency score, VRS | **1.15** | **1.29** | **−0.14** | **−11.00%** |
| Average efficiency score, NIRS | **1.16** | **1.36** | **−0.19** | **−14.29%** |
| Average efficiency score, SE | **1.02** | **1.10** | **−0.07** | **−6.81%** |

Out of the six DEA models generated in the Step 2 we have arbitrarily chosen the model *DMU_Goal* = "Input Oriented" and *DMU_Criterion* = "CRS" to illustrate Step 3 of our methodology.

## Step 3: Describe the Relative Efficiency Categories

### *Description of Step 3*

In this step, we will use DT induction to generate rules that can describe the relative efficiency categories in terms of the input and output variables of the DEA models. This will require the inclusion of a target variable (say "*EfficiencyCategory*") that identifies the efficiency category. For the case of the homogeneous sample of DMUs, we would use a binary variable to indicate whether the DMU is relatively efficient. For a homogenous data set, we have to use a categorical variable that allows for two efficiency categories per cluster. Thus, in the case if the CA in the Step 1 resulted in the solution with two clusters, the domain of values for our categorical target variable could be represented as follows:

"11" – Relatively Efficient DMU with membership in the Cluster 1,
"10" – Relatively Inefficient DMU with membership in the Cluster 1,
"21" – Relatively Efficient DMU with membership in the Cluster 2,
"20" – Relatively Inefficient DMU with membership in the Cluster 2.

If we do not take into consideration heterogeneity of the sample, then we end up with the following domain of values:

"1" – Relatively Efficient DMU in the sample,
"2" – Relatively Inefficient DMU in the sample.

In general, we will assume that our target variable (say *EfficiencyCategory*) is a concatenation of the relevant cluster identifier attribute *ClusterNum* and relative efficiency status attribute EfficiencyStatus. Once the data set has been amended to include the target variable *EfficiencyCategory*, DT induction is used to generate a DT that can be used to describe the efficiency categories.

*Illustration of Step 3*

For our illustration of this step, since the result of Step 1 indicated that our data set was heterogeneous with two groups, we first populated our target variable "Cluster Efficiency" with the following values:

- Value of "21" was assigned to the "*Efficient Leaders*", those TEs that belong to the "leaders" cluster and were assigned the score of "1" by DEA
- Value of "20" was assigned to the "*Inefficient Leaders*", those TEs that belong to the "leaders" cluster and were assigned the score of less than "1" by DEA
- Value of "11" was assigned to the "efficient majority", those TEs that belong to the "majority" cluster and were assigned the score of "1" by DEA
- Value of "10" was assigned to the "*Inefficient Majority*", those TEs that belong to the "*Majority*" cluster and were assigned the score of less than "1" by DEA

The number (*N*) of DMUs in each of our four categories are: 38 *Efficient Leaders*, 31 *Inefficient Leaders*, 43 *Efficient Majority*, and 68 *Inefficient Majority*.

We then generated our DT model that enabled us to identify conditions associated with our four categories. In Table 26.6 we display a sub-set of the rules that are associated with this DT. For each category, we selected a pair of rules, each of which had a strong probability (i.e., **Prob** > 0.90) for the occurrence of the associated category given the condition component of the rule.

**TABLE 26.6**
**Pairs of Rules that Describe the Efficiency Categories**

| Condition | Efficiency Category: Group | Efficiency Category: Efficiency Status | N | Prob |
|---|---|---|---|---|
| *Productivity Ratio per Telecom Worker ≥ 4.1754445351 & Annual Telecom Investment per Worker ≥ $58* | *Leader* | Efficient | 14 | 1.00 |
| *Total Telecom Services Revenue per person ≥ $210 & Full-Time Telecommunication Staff % ≥ 0.0039016912 & Productivity Ratio per Telecom Worker < 4.1754445351 & Annual Telecom Investment per Worker ≥ $58* | *Leader* | Inefficient | 5 | 1.00 |
| *Full-Time Telecommunication Staff % < 0.0039016912 & Productivity Ratio per Telecom Worker < 4.1754445351 & Annual Telecom Investment per Worker ≥ $58* | *Leader* | Inefficient | 11 | 1.00 |
| *Full-Time Telecommunication Staff % < 0.0031414015 & Productivity Ratio per Telecom Worker ≥ 3.8043909395 & GDP per Capita ≥ $519 & Annual Telecom Investment per Worker < $33* | *Majority* | Efficient | 8 | 1.00 |

| Condition | Efficiency Category: Group | Efficiency Category: Efficiency Status | N | Prob |
|---|---|---|---|---|
| *Total Telecom Services Revenue ≥ 0.0118204323 & GDP per Capita < $519 & Annual Telecom Investment per Worker < $33* | *Majority* | Efficient | 22 | 1.00 |
| *Productivity Ratio per Telecom Worker < 3.8043909395 & GDP per Capita ≥ $519 & Annual Telecom Investment per Worker < $33* | *Majority* | Inefficient | 39 | 1.00 |
| *Full-Time Telecommunication Staff % ≥ 0.0031414015 & Full-Time Telecommunication Staff %< 0.0054371357 & Productivity Ratio per Telecom Worker ≤3.8043909395 & GDP per Capita < $519 & Annual Telecom Investment per Worker < $33* | *Majority* | Inefficient | 12 | 0.92 |
| *Productivity Ratio per Telecom Worker < 2.002357802 & $33 ≤ Annual Telecom Investment per Worker < $58* | *Majority* | Inefficient | 6 | 1.00 |

## CONCLUSION

DEA is a good data analytic tool discriminatory power of which, however, is somewhat dependent on the homogeneity of the sample of DMUs. Relative efficiencies of the NFL players, for example, could not be meaningfully compared unless the position played, which could be dependent on height, speed, and weight, is taken into consideration. Similarly, relative efficiencies of the designated hitters and catchers of MLB in terms of the production of the home runs should not be compared directly either. Thus, necessity arises to find a way to conduct DEA while taking into consideration some of the important differences between the groups of the DMUs in the sample.

In the case of this study, we have identified two sub groups within our sample of 18 TEs, the "*Majority*" and the "*Leaders*". Having this insight in mind, we have three options of conducting DEA. First, we could disregard this information and proceed directly with DEA. Second, we may conduct separate DEA per each cluster. Third, we have an option of proceeding according to the proposed in this chapter methodology.

We would like to argue that our methodology allows for achieving of a better discriminatory power of DEA than its two alternatives. In order to do so, we offer a comparison of the pairs of rules that were obtained by utilizing each of the three mentioned above options. In order to avoid repeating contents of Table 26.6 in this section of the chapter, we provide only the sets of the decision rules corresponding to two other options.

1st option: **DT based on DEA only** ("efficiency" encoded as "1" – efficient, "0" – inefficient), 57 efficient DMUs, 123 inefficient DMUs

The results of this DEA demonstrate that a number of transitional economies have obtained a rating of being relatively efficient. It does not mean, however, that all of the countries that were deemed relatively efficient are in fact efficient. A common characteristic of DEA models is that they tend to evaluate as efficient those DMUs that have the smallest input values, or, the DMUs with the largest outputs (Ali, 1994). Thus, the approach consistent with the 1st option does not allow an investigator to determine whether the relative efficiency of a DMU is caused indeed by its efficiency, or whether a DMU was awarded a relatively efficient status because it had the smallest level of the inputs in the sample.

Consequently, one of the shortcomings of the first DT model is that it does not allow an investigator to incorporate the results of the CA, i.e., an "*Efficient Leader*" is highly likely to be very different from an "*Efficient Majority*".

2nd option: **DT based on DEA, one DT model per cluster**

a. "*Leaders*" cluster ("efficiency" encoded as "1" – efficient, "0" – inefficient), 38 efficient DMUs, 31 inefficient DMUs
b. "*Majority*" cluster ("efficiency" encoded as "1" – efficient, "0" – inefficient), 43 efficient DMUs, 68 inefficient DMUs

These models, consistent with the 2nd option, improve on the first model by being cluster-specific. Let us recall, nevertheless, that while some of the TEs are "permanent residents" of one cluster, the other TEs are "migrants", for they change their membership depending on the year. Consequently, none of the generated by this approach models would help us to inquire into the question why, for example, Lithuania was a member of the "*Majority*" cluster for the period from 1993 to 1998, but became a member of the "*Leaders*" in the period from 1999 to 2002.

Considering the presented above alternatives, it would appear that the results of our methodology, presented in Table 26.6, allow for achieving of a higher discriminatory power of DEA. While the results of the conventional DEA only yield the efficiency scores for each DMU in the sample, the results of our approach yield the efficiency scores and, in the case of the heterogeneous domain, the membership within the subset of the sample for each DMU. This (considering that the most decision makers are interested not so much in learning whether or not a given DMU is inefficient, but rather why it is inefficient) allows for a higher degree of granularity of the subsequent analysis, as it was demonstrated by comparing Tables 26.6, 26.7, 26.8, and 26.9.

Another benefit of our methodology is that it takes an approach of an "external augmentation" of DEA, meaning, it does not require any changes to or alterations of DEA itself. As a result, our methodology is not model-specific and, consequently, could be applied to any DEA model. However, despite the contributions that this research makes, we must acknowledge that our study is not without its limitations.

First limitation of this research is associated with the use of CA. At this point, we cannot offer strict criteria determining whether the sample should be considered homogeneous or heterogeneous. Thus, despite the explicit nature of testing for heterogeneity by means of CA, the determining decision regarding the sample still lies with the decision maker.

Second limitation of our study is associated with the use of DT induction. At this point, we cannot suggest to a decision maker what splitting criteria and what settings yield a better tree. As a result, this issue as well resides in the domain of the responsibilities of the decision maker.

**TABLE 26.7**
**First Option: Pairs of Rules that Describe the Efficiency Categories**

| Condition | Efficiency Category: Group | Efficiency Status | N | Prob |
|---|---|---|---|---|
| *Full-time telecommunication staff % < 0.0041924905 & Productivity ratio per telecom worker ≤ 5.4191734889* | *Whole Set* | Inefficient | 25 | 0.92 |
| *Total telecom services revenue ≤ \$51,715,239 & Total telecom services revenue per telecom worker < \$60,794 & Productivity ratio per telecom worker < 5.4191734889* | *Whole Set* | Inefficient | 91 | 0.956 |

**TABLE 26.8**
**Second Option: Pairs of Rules that Describe the Efficiency Categories in "Leaders" Cluster**

| Condition | Efficiency Category: Group | Efficiency Status | N | Prob |
|---|---|---|---|---|
| *Total telecom service revenue per capita ≤\$101 & Productivity ratio per telecom worker < 4.1754445351 & Annual telecom investment % GDP < 0.0166 & Full-time telecommunication staff % ≤0.0043712305* | *Leaders* | Efficient | 12 | 1.0 |
| *Productivity ratio per telecom worker ≥ 4.1754445351* | *Leaders* | Efficient | 17 | 1.0 |
| *Full-time telecommunication staff % < 0.0043712305 & Productivity ratio per telecom worker < 4.1754445351* | *Leaders* | Inefficient | 21 | 0.905 |

**TABLE 26.9**
**Second Option: Pairs of Rules that Describe the Efficiency Categories in "Majority" Cluster**

| Condition | Efficiency Category: Group | Efficiency Status | N | Prob |
|---|---|---|---|---|
| *Productivity ratio per telecom worker < 3.8106041166 & GDP per capita ≤ \$519* | *Majority* | Inefficient | 47 | 1.0 |
| *Total telecom services revenue % ≤0.0118204323 & GDP per capita < \$519* | *Majority* | Efficient | 22 | 1.0 |

## ACKNOWLEDGMENT

Material in this chapter previously appeared in: Increasing the discriminatory power of DEA in the presence of the sample heterogeneity with cluster analysis and decision trees. *Expert Systems with Applications*, *34*(2), 1568–1581.

## REFERENCES

Ali, A.I. (1994). "Computational aspects of DEA." In: Charnes, A., Cooper, W.W., Lewin, A., and Seiford, L.M. (eds.), *Data Envelopment Analysis: Theory, Methodology and Applications*. Kluwer Academic Publishers, Boston, MA, pp. 63–88.

Charnes, A., Cooper, W.W., & Rhodes, E. (1978). Measuring the efficiency of decision making units. European Journal of Operational Research, 2, 429–444.

Charnes, A., Cooper, W.W., Lewin, A.Y., and Seiford, L.M. (1994). *Data Envelopment Analysis: Theory, Methodology and Applications*. Kluwer Academic Publishers, Norwell, MA.

Hong, H., Ha, S., Shin, C., Park, S., and Kim, S. (1999). "Evaluating the efficiency of system integration projects using data envelopment analysis (DEA) and machine learning", *Expert Systems with Applications*, Vol. 16, pp. 283–296.

Seol, H., Choi, J., Park, G., and Park, Y. (2007). "A framework for benchmarking service process using data envelopment analysis and decision tree", Expert Systems with Applications, Vol. 32, No. 2, pp. 432–440.

Sohn, S., and Moon, T. (2004). "Decision tree based on data envelopment analysis for effective technology commercialization", *Expert Systems with Applications*, Vol. 26, No. 2, pp. 279–284.

# 27 An Exploration of the Intrinsic Negative Socio-Economic Implications of ICT Interventions

## INTRODUCTION

The socio-economic impacts of information and communication technology (ICT) have been extensively researched in diverse contexts, within different time frames, and by various methods of inquiry. It is fair to say there is a general consensus that investments in ICT result in "good things" of various kinds. After reading findings of published research, one feels that investing in ICT is similar to buying a night-time cough medicine to help your cold feel better next morning. You pay the price (for ICT does not come free), you take the medicine (ICT needs to be integrated and implemented), and you wait (there is a time lag). And it usually works as advertised – you do feel better and all seems to be well. The medicine does not work perfectly, but it never completely fails. So, overall, this medicine, as well as ICT, is a "good thing". However, a medicine always comes with a warning of side effects – there is a disclosure mentioning an additional price to pay for getting a "good thing" going.

Unlike medicine, side effects of "good things" brought about by ICT are almost impolite to mention – for the technological progress and productivity growth usually come first and rank high on any economy's list of priorities. And they should come first, for it is better to make money, generate wealth, and then re-distribute them if needed, than to be ready to re-distribute, but having nothing to give. But there is a price to pay for economic development – side effects are there whether we want it or not.

Technology and progress are like a big ship taking off the pier and heading full speed to the sea – moving economies forward, but leaving some small boats greatly disturbed in its wake. We should not try to stop, or slow down, this big ship, but we should be aware of what it leaves in its wake, so we are prepared for little boats to be in trouble.

This is, really, a matter of comparing and contrasting "big deals". Increasing efficiency and effectiveness of a business process, or a governmental service, or an Internet access is a big deal, indeed. We can trace the impact to growth in productivity, and then, in turn, to the macro-economic outcome – growth in GDP. But there are other "bid deals".

Big deal if a local mom-and-pop store went out of business – are you not happy, instead, to use a self-checkout station in a local supermarket? Well, a cashier that knew your name is gone, but the process is much quicker now, no cashier is needed and prices are lower. Much lower than in that mom-and-pop store, now out of business, where locals used to meet.

Some of the negatives are situation-specific and unique to a given setting offering researchers an anecdotal evidence. Other negatives are philosophically inevitable – they will be present regardless of the context. If you are in business of designing bombs, then you should aim for making the most efficient and effective once – just business, nothing personal. However, it is good to be aware of a collateral damage of a bomb maker' success.

Our objective in this editorial is to bring attention of our colleagues to some of the "negative inevitables" of positive impacts of ICT (Bosamia, 2013; Tarafdar et al., 2013, 2015). Along the way we will try to suggest a way of categorizing the negatives. Finally, we will attempt to propose possible routes of investigating them. A full disclosure to our readers – it is not our purpose to rain on a parade of positive impacts of ICT, but it is our goal to suggest to our colleagues that ICT-driven initiatives aiming for "good things" should come with a warning label mentioning side effects.

## SOCIO-ECONOMIC IMPACT OF ICT

### Tools, Machines, and ICT

Before ICT was a "good thing", there were machines, and before machines there were tools – "good things", all. It is important to mention them because they are inextricably linked together.

An introduction of tools, machines, and ICT has a shared objective – to make an execution of work more effective and efficient, which pursued according to the same strategy, that of elimination and substitution. But while tools, machines, and ICT share a common objective and a common strategy, they implement them by following different routes. Let us spend a minute clarifying terminology used in this chapter.

Physical work is an act of applying energy to matter with the purpose of transforming its properties. Physical work has its counterpart in digital domain. Digital work is as an act of collecting, manipulating, storing, or disseminating of data – an act of applying energy to data with the purpose of changing its properties – location, format, encoding, encryption, etc. A work element is a part of work, a task, that could be performed by a single worker. Sharpening of a knife or an entry of a record into a database are examples of work elements.

A tool is an object, natural or manmade, allowing for a more efficient and effective performance of a function within a work element. A knife sharpener and a document scanner are examples of tools.

A machine is a system incorporating a tool(s) allowing for a more efficient and effective performance of work. A knife sharpening machine is an example of a machine in the mechanical domain. A digital graphing calculator is an example of a digital machine.

An Information Technology (IT) is an electronic symbol generating and processing system comprising software and hardware that allows for assessing, via generating information, the working order, or a state, of a machine. IT-enabled sharpening machine allows for monitoring a number of surfaces sharpened, a wearing out of the sharpening tool, and the time-to-service left – once it is up the warning message appears. A laptop is an example of IT that controls the state of multiple virtual machines – applications.

ICT is an IT-based system that creates new, or optimizes existing, information channels which allow for controlling a set of virtual or IT-controlled machines involved in a complex workflow (work processes). An automated order fulfillment center and a cloud-based online payment system are examples of ICT.

## Routes of Elimination and Substitution

Let us take a look at various ways by which tools, machines, and ICT bring about "good things". A tool allows for increasing efficiency and effectiveness of performing a work element via eliminating a direct involvement of a human body part(s) in performing work by means of substituting it with an object. Hitting your enemy in the head using a hammer is more efficient and effective than using a fist. Digitally recording a song is more effective and efficient than doing so in analog way, whether recording it on a vinyl disk or via transcribing its text and notes.

A machine allows for increasing efficiency and effectiveness of work via mechanization or automation – by eliminating human involvement in repetitive work by means of substituting humans equipped with tools. IT allows for increasing efficiency and effectiveness of work via substituting humans controlling the machines. An automated production line' control system, or a self-checkout counter at the grocery store both eliminate significant human involvement.

ICT takes it a step further and achieves the objective via substituting human decision-making process geared toward an optimization of a workflow. Just-in-time (JIT) manufacturing system or Apple' iTunes store are examples here. In both cases human involvement is minimized.

An artificial intelligence (AI)-enabled ICT aims to replace human decision making within a novel context (e.g., changed business environment) of a workflow. Collaterally, AI is a tool that attempts to manipulate human decision-making processes.

The impact of elimination and substitution via tools, machines, and ICT is compound because the elements in the elimination and substitution chain are interdependent. Better tools allow for integrating them in more complex machines that are controlled by the increasingly sophisticated software component of IT, and the complex machines could be interconnected via ICT (especially AI-enabled ICT) to comprise a complex workflow. Consequently, if entity A invests in a more sophisticated ICT than entity B, then, ceteris paribus, entity A should be able achieve a greater level of effectiveness and efficiency of work via elimination and substitution alone.

However, there is inevitable impedance mismatch within the elimination and substitution chain if the chain incorporates analog physical work and digital symbolic processing and control. This means that even the most sophisticated warehouse or distribution center (e.g., Amazon' ones) will not be able to eliminate as many people as would be possible otherwise. The maximum effect the chain achieves in the case of moving all the work into a digital domain (e.g., Instagram replaced 100K Kodak' employees). For example, a branch of a bank with an on-ground presence will have to have some on-ground employees left as long as they deal with physical money. If, however, the bank or a financial institution moves its operations completely into a digital domain, then the success of the substitution and elimination chain could be much greater.

In any case, digital or analog, any implementation of ICT will result in inevitable elimination and substitution, somewhere along the chain, of human involvement.

Furthermore, this impact remains regardless the type of workflow – whether the work is accomplished for the benefits of shareholders (e.g., any implementation of Enterprise Systems), or if it is done for personal, leisurely purposes (e.g., subscribing to a new service or downloading a new app to your smartphone). At this point we are in a good position to state Negative Inevitable #1:

*Investments in ICT will result in a substitution and elimination of human involvement.*

## Conditions for Elimination and Substitution

In principle, there are two conditions when technology (e.g., tools, machines, ICT) will not be used to substitute and eliminate human involvement. First, when there is no available technology to do so and, second, when technology is available, but it is cheaper to use humans.

There are some important implications resulting from these conditions.

In the first case it is worth considering a situation under which "available technology" is equivalent to "affordable technology", for both cases will result in the same outcome – there will be no substitution. In the instance of ICT this is relevant, for a competitive advantage nowadays comes not from an ability to do work (which is a given), but from an ability to coordinate complex workflows. And it is in the domain of coordination of work that ICT shines. Thus, by improving the chain "resource → tools → machines → IT-enabled machines" an economy that cannot afford ICT may only go so far. This means that an economy that cannot afford needed ICT will not be able to make more money to afford it, but will have to wait until prices of such ICT are reduced sufficiently to be affordable. Basically, dragging further and further behind those who can afford it now.

In the second case, if we ask ourselves a question "Where is it cheaper to use humans than technology in 21st century?" we can come up with a surprisingly long list of places. In all of those contexts' investments in ICT, however important, should not be considered an overriding priority. Instead, the priority should be in transitioning from human labor to machines, and then to IT-enabled machines.

But, for many economies that cannot afford ICT there are external sources that could help with the funding. So, what would be a negative consequence of getting a technology that an economy cannot really afford or could do without because human labor is cheaper? To answer this question, we need to consider the complete chain "*Resources → Tools → Machines → IT-enabled Machines → ICT*" to see that the introduction of ICT will expose shortcomings and conflicts within the next link below – "IT-enabled machines". Basically, an introduction of advanced (for a given context) ICT will result in a situation of "OK, now we have the technology to connect IT-controlled machines, but how do we get the machines up to speed now?" At this point we can state Negative Inevitable #2:

*Introduction of advanced-for-the-context ICT will magnify the existing, and identify new, deficiencies within the state of IT-enabled machines.*

## Pragmatics and Ethics of Implementation

A different set of potential problems emerges if we consider the inevitable gap between an *output* of the implementation of ICT and the *outcome* of it. Investments in ICT are allocated primarily towards impacting the current infrastructure of the context and, naturally, the intended result is delivered via the implemented ICT, which is not an end to itself, but a mean to a greater end – a positive socio-economic impact. Overall, the following chain of links is traceable: "*Investments in ICT → Implementation of ICT → Impact of Implemented ICT*", which is, for all intents and purposes, a chain "*Investments in ICT → Output → Outcome*".

Thus, the actual resultant ICT is the *output* of the process of implementation of investments in ICT. It is of no use to have investments in ICT available as an end game, for it is just a line on a balance sheet that makes an account look good, but doing not much beyond that. Consequently, investments must be allocated and implemented to produce an impact, and the most common way to introduce new technologies is via using ISD methodologies.

The overwhelming majority of methodologies used in the process are functionalist – based on mechanistic approach and driven by pragmatic considerations. The quality of the output, therefore, is judged based on the appropriate to functionalism criteria, which, in the case of the chain "*Resource → Tool → Machine → IT-controlled Machine → ICT*", translates into increased effectiveness and efficiency of the chain.

A greater impact of ICT, however, resides not in the output of the process of implementation, but in the outcome of the process within a larger system. This forces us to move out of a narrow technological domain into a more complex social domain, where the assessment of the impact of ICT is no longer driven by pragmatics, but, instead, by ethics.

Resultantly, it is only expected to anticipate a presence of incongruence between the output of ICT and the outcome of it, for pragmatics and ethics are based on different principles and assumptions. Now we are well-equipped to state Negative Inevitable #3:

*An assessment of the impact of ICT will require reconciliation of differences between the pragmatics of the implementation of ICT and the ethics of the expected results of the implementation.*

Let us spend some time considering the aspects of a social domain that could be impacted, positively and negatively, by ICT.

## Dimensions of Social Impact of ICT

Introduction of ICT results in changes in the host's social environment – even if the implementation itself is purely technical, impacts of the implementation spread beyond the boundaries of purely technological domain. Regardless of the context, no society is homogeneous, and we need to consider what sort of dimensions are responsible for heterogeneity, for every one of them will be impacted by ICT.

We will consider only the most obvious ones – economic, cultural, political, and religious. At this point we are not going to be concerned with the "interaction effects" such as "wealthy liberal", "poor conservative", and so on. An uber dimension that is responsible for differentiating a society is economics. In the case of introducing a new and improved ICT the financially better off groups of society will disproportionally benefit from the change. After all, it is the members of any society that have more advanced IT that benefit from ICT the most. Consider having a new high-speed highway – the introduction of a new road is a good thing for any driver, but the drivers having better cars will benefit the most, while the owners of the old ones may struggle to adjust their vehicles to new conditions (e.g., higher speed limit).

At this point Negative Inevitable #4 seems to be in order:

> *Economically disadvantaged groups of a society will be negatively impacted by the introduction of new ICTs due to the pressures of the compliance with the new technology and the limited use of new capabilities due to restrictions associated with the old technology.*

However, pressures on a technological front are only a beginning of the impact of new ICT – let us consider other ones as well.

## Platform, Message, and Target

It is fair to say that almost any coherent, organized, and recognizable/identifiable sub-group in a society has its own platform and its own message, as well as an intended target of the message. We define a *platform* as a collection of the available information channels allowing for transmitting of the group's message. A *message* is a statement containing a set of propositions reflecting the political, cultural, or religious values (or a combination of them) that the group wants to impart to a target. A *target* is an intended audience of the message, which could be another sub-group, or a society in general. Introduction of new ICT brings some interesting consequences impacting platforms, messages, and targets that are worth considering.

A new ICT will allow for creating new and optimizing existing information channels. Thus, in the case of an affluent sub-group, its platform will become *wider* due to quantitative and qualitative changes the new ICT will bring. *Quantitative changes* are associated with the platform gaining a greater number of information channels of *the same type*, and *qualitative changes* are associated with gaining a greater *variety of the types* of information channels (e.g., text, audio, video, and digital art).

The platform will also become *taller* via allowing for reaching a wider audience – a bigger target group. So, a new ICT is a "good thing" for a well-off group.

In the case of an economically disadvantaged sub-group, however, the platform will become *relatively narrower*, because it will require a newer technology to utilize new ICT to the full extent. Also, the platform will become *relatively lower*, because some of the previously accessible audiences will transition to using new technology that is not available to the poorer sub-group.

It is possible, of course, that the new ICT will not result in reduction of the width and height of the platform of an economically disadvantaged group. However, because the economically advantaged group will take an opportunity of the new ICT, this will make their platform wider and taller than it previously was. Hence our use of the term *relatively* – even if the platform of the disadvantaged group stays the same, its width and height deteriorates *relative* to the competition. Consequently, Negative Inevitable #5 is as follows:

> *Introduction of new ICT will result in a relative deterioration of the social platform of the economically disadvantaged subgroups of a society.*

Some interesting things happen to messages. While the substance, the content of the message may stay the same (it does not have to, but let us assume the simplest scenario), the two characteristics of a message will be impacted greatly. These characteristics are *format* and *intensity*. New ICTs offers new types of information channels, which allow for a delivery of a message in a greater variety of formats (e.g., text, audio, and video) and with a greater frequency. Increasing a variety of formats and a frequency of delivery neither come free, nor cheap. Hence, an introduction of new ICT will result in a *relative decrease of intensity* and a relative *decrease of attractiveness of the format* of the message in the case of economically disadvantaged sub-groups. It is important to note that we are not talking about a *content* – the substance of the message, but only about its' packaging and a frequency of delivery.

Resultantly, we put forward Negative Inevitable #6 as follows:

> *For economically disadvantaged subgroups an introduction of new ICT will bring a relative decrease in (1) frequency of delivery* and *(2) attractiveness of the format* of their message.

Introduction of a new ICT also impacts targets. It is safe to say that the majority of active sub-groups in any society aim to reach outside its own constituency to promote their set of values. Let us consider a simple scenario where a sub-group has its loyal following and constituency that will not be impacted by the introduction of new ICT. The outside targets, however, will be impacted by the new ICT – the impact is based on what we call continuously rising expectation regarding the *wrap* and *drop*. It is only to be expected that people expect messages to be nicely *wrapped* – delivered to them in the best possible/available format (e.g., high-production video vs. text), and continuously *dropped* – delivered to them regularly and often. Targets, whether they want it or not, will become fickler as they are offered more choices which raise their expectations.

It is only reasonable to suggest, then, that *format wars* and *delivery battles* will take place, and will play out with fairly predictable results. Economically better off sub-groups will win format wars because they have a financial edge in producing, or paying to produce, messages in a *better wrapper* – in a higher quality format. Also, it is only to be expected that more affluent sub-groups will win delivery battles because they will have more resources for delivering messages more frequently and on-demand – a *fast food delivery model* will prevail in digital domain.

Simply put, a Darwinian' *message selection by a target* plays out, where fast and pretty survives, and slow and substantive dies.

Now we can state Negative Inevitable #7, as follows:

*For economically disadvantaged subgroups an introduction of new ICTs will make reaching the target of their message relatively harder.*

Let us now consider some political and cultural implications of ICT, where two negatives come to mind. First, an introduction of an advanced ICT will result in a further consolidation of political powers around the affluent centers. The reason is simple – the fundamental architecture of the Internet is based on a client-server model, and the centers of control and influence are usually formed around the servers. Consequently, if the resources of an economy are in the hands of a government (or, government – controlling entities), then mechanisms of political suppression will become more effective and efficient, specifically in the context of non-democratic societies. Simply put, bigger, better, faster servers will force re-centralization of decentralized model of the Internet-based computation, with all the political impacts it may bring. We suggest a Negative Inevitable #8, as follows:

*In the societies of a globalized world introduction of new ICT will strengthen the existing political status quo.*

A second implication is associated with the purpose of the culture of a society, which is to provide a shared, stable and consistent world-view to its members. An introduction of such change agent as advanced ICT may not result in a significant cultural impact in the context of the Western world, but in a context of societies that are based on different cultural premises it may. We should not forget that ICT is a messenger of such powerful mechanistic ideas as positivism, determinism, and individualism, and it requires a compliance with them to be utilized effectively and efficiently. It is not entirely unlike the democratic ideas of the West, which were successfully adopted by many post-Soviet countries in Europe, but did not root well in Arab world and Africa due to fundamental cultural differences. Simply put, ICT requires an appropriate cultural soil to take off, for, otherwise, cultural clashes are inevitable. We offer a Negative Inevitable #9, as follows:

*An expected corollary to the outcome of the implementation of the ICT in the non-Western context of the globalized world is a struggle to reconcile the values of the Western culture with the local one.*

Additionally, there are also ICT-driven changes that impact inter-economy situation. We consider them next.

## Competing with Others: Additional Implications

An introduction of the new ICT will also have an inter-economy impact, for changes brought about by ICT rarely can be contained within a single economy. Let us consider two simple scenarios to illustrate the implications.

The first scenario refers to the situation when Economy A wants to gain a competitive edge over Economy B via the introduction of the new ICT. We consider such situation playing out via two intuitive cases.

The first case is when both economies have well-integrated and developed chains "*Resources → Tools → Machines → IT-enabled Machines*" and investing in new ICT does make a competitive sense for Economy A. However, the resultant competitive edge will be short-lived, for all it takes for Economy B to catch up is to invest in its own new ICT as well. Thus, once Economy B also gets its new ICT, the obvious next step for Economy A is to make the complete chain "*Resources → Tools → Machines → IT-enabled Machines → ICT*" leaner relative to Economy B. And this "leaning out" of a workflow chain will impact the involved labor force – *elimination and substitution* of the workforce will take place. However, the impact on the labor force will be greater because it will be driven by two factors working in synergy. The first factor is new ICT and its impact on a workflow chain, where "as-is" state with X employees will be replaced by "to-be" state with X-n employees.

The second, compounding factor, is a competition with Economy B. Given the fact that Economy B will go through its own "as-is" to "to-be" transformation resulting in Y-m employees, Economy A will attempt to gain a competitive advantage via the second round of elimination and substitution aiming at prevailing over Economy B via assuring that the equation $(X-n) < (Y-m)$ remains valid.

The second scenario refers to the situation when Economy A, in the absence of the well-integrated and developed chain "*Resources → Tools → Machines → IT-enabled Machines*", invests in new ICT aiming to attract outside investments (e.g., FDI) and, resultantly, gain an edge over Economy B via this route. However, it is only reasonable to assume that such plan will fail – two reasons come to mind. First, no external funds may come about if the outside investments will aim to take advantage of well-developed and integrated chain "*Resources → Tools → Machines → IT-enabled Machines → ICT*". Second, external funds could be allocated under condition of improving the "*Resources → Tools → Machines → IT-enabled Machines → ICT*" chain. Because investments more often than not are allocated for the purpose of making profit, external entities will have a goal of using Economy A to compete against Economy X. This results in a familiar scenario of elimination and substitution resulting, first, in X-n as a result of new ICT, and second, (X-n)-k as a result of the pressure from external investors competing against Economy X. At this point we put forward Negative Inevitable #8, as follows:

*An introduction of new ICT as a tool of inter-economy competition will result in an increased level of elimination and substitution of the workforce.*

## Competing with Others: Social Implications

Inter-economy competition would also bring about some interesting political, cultural, and religious implications. While the *extent* of the impact along those dimensions is context-specific and indicative of local inflections, the general tendency can be traced. ICT-based competition will result in economic stratification of a society based on three consequences of the implementation of ICT. First, there will be displaced employees whose jobs were eliminated and substituted by IT. This is important to note: this is not a scenario of "you can get a raise or you can get a cut", but, rather, it is a scenario of "sorry, there is nothing for you". Displaced employees will come from all three organizational levels – operational, tactical, and strategic.

Second, there will be a drive to eliminate middle managers – some of them will be eliminated, others will become a part of a strategic level, and yet others will descend to operational level. Appropriate levels of compensation (e.g., increased, decreased, and welfare) will determine the strata of the destination.

Finally, there will be a greater disconnect between operational level of a firm, and its strategic level. The reason is simple – the interface, the middle managers, is gone.

This stratification of the society based on economics results in strengthening of the cohesiveness of the political, cultural, and religious sub-groups. A general need for an affiliation will drive people towards joining or becoming more active in, what is perceived as, "stable, consistent, and reliable" social group. That, in turn, will lead to "us vs. them" attitude of the members of sub-groups. Interestingly, this attitude will have two directions – internal and external. The internal direction will be pointed against the sub-groups of the homeland economy that are perceived to be responsible for negative changes that took place and impacted the local labor force. The external direction will have its aim on the members of "other" economy(-ies) that are perceived to be responsible for the deterioration of the economic situation in the homeland.

This allows us to put forward Negative Inevitable #9, as follows:

*An introduction of new ICT as a tool of inter-economy competition will result in a compartmentalization of the social environment leading to the increased levels of hostility towards local and external groups perceived to be responsible for the deteriorating economic conditions of negatively impacted subgroups.*

## Impact of Collaboration

Finally, let us consider a case of an inter-economy collaboration – the case of two or more countries working on an ICT project. Unless the economies involved are fairly aligned, from a political, social, and cultural standpoint (e.g., Finland collaborating with Sweden), certain domination-based scenarios are bound to unfold. Culturally, a clash is inevitable and two simple outcomes of such clash are a disintegration of a project, or a submission of a weaker partner to the dominating counterpart. Politically, the impact will be felt by the "junior" member(s) of the project team because the power structure of the weaker member of a team will be under constant pressure of a "stronger" partner. Socially, the impact will be felt if the assignment of the roles and responsibilities of a weaker partner does not match the model of a more powerful entity. At this point we can state another Negative Inevitable, #10, as:

*International collaborations on ICT will face the conflicts associated with and brought by the social, political, and cultural diversity of the collaborating members that will be resolved via the domination of more powerful member of the group.*

## Investigating Negative Implications of ICT: What Is the Plan?

At this point it is time to start contemplating the means by which *Negative Inevitables* mentioned in this editorial, as well as additional ones yet to be discovered and argued for, could be investigated.

We would like to propose a simple, two-step plan of action. First, researchers working in the area relevant to the subject matter must take into consideration, explicitly, negative implications of ICT. And this is not an easy thing to accomplish, for one would have to account for the negatives that take a part of ICT-based glory away. It will, indeed, require a paradigm shift away from the all too familiar "IT is good for development" attitude. Is a sharp knife good? Well, it depends … So, this "it depends" perspective must be embraced.

This allows us to formulate Research Suggestion #1:

*An investigation of positive socio-economic impacts of ICT should also consider, explicitly, collateral negative outcomes.*

Second, investigators must consider the appropriate targets belonging to the different levels of inquiry.

We consider three levels of granularity at which an investigation can be conducted – case studies, frameworks, and grand theories. In our view the three levels relate to each other as follows. Grand theories espouse relationships between general, context-independent constructs, frameworks operationalize the constructs in the context-specific manner, and case studies draw rich pictures of the context from "boots on the ground" perspective. Ideally, there will be a hierarchy and continuity of representation of the results and the data between the levels. Thus, findings of a case study should be possible to frame as a representation of a context-specific construct, and a context-specific construct should be possible to abstract and represent as a more general context-independent construct of a grand theory.

If one is to undertake a case study within the context of a single economy, then the local inflections must be identified. Meaning, it is fine and well that the farmers now have the cell phones to get the latest and the most accurate prices for their produce, but what happened to the dealers who used to direct the flow of the goods to the appropriate locations?

This allows us to formulate Research Suggestion #2:

*A case study dedicated to investigating socio-economic impacts of ICT should identify local, specific to the context of the case study, collateral negative outcomes.*

In the case of an investigation relying on a framework, a researcher should be able to suggest, first, how a context-specific construct could be generalized to a context-independent one of a grand theory, and, second, how a context-specific construct may manifest itself in the "boots on the ground" setting. For example, if the results of a case study showed that an introduction of cell phones to a farming community resulted in elimination and substitution of local produce dealers, then what would the appropriate framework-level construct be?

We would like to formulate Research Suggestion #3, as follows:

*A framework-based research should incorporate within its model a context-specific construct representing collateral negative outcomes of ICT such that: (1) the construct could represent more specific local inflections of a case study-context, and, (2) the construct could be generalized to the level of theory.*

Finally, if conducting a study backed by a theory, the investigator should attempt to include a general construct representing a collateral negative impact of ICT. Clearly, creating such construct will be driven more by philosophical consideration, then by practice. Nevertheless, the investigator must keep in mind that such a construct should be translatable to the level of a framework, and then to the level of a case study.

At this point we can formulate our Research Suggestion #4:

*A theory-based investigation should incorporate a context-independent construct representing collateral negative outcomes of ICT based on, and resulted from, the substitution and elimination nature of ICT.*

We provide a brief illustration of the suggestions in Table 27.1.

**TABLE 27.1**
**Research Suggestions Taking into Consideration Negative Impacts of ICT**

| Level | Neoclassical Growth Accounting | Modified Neoclassical Growth Accounting |
|---|---|---|
| **Theory** | EconOutput = (Labor, Capital, TFP) | EconOutput = (Labor, Capital, TFP, ***-DisplacedLabor***) |
| **Framework** | GDP = (IT_Labor, IT_Capital, TFP) | GDP = (IT_Labor, IT_Capital, TFP, ***-Displaced_IT_Labor***) |
| **Case Study** | Cell phone usage by farmers in Economy X resulted in increasing level of income due to the improved access to a local pricing system | Cell phone usage by farmers in Economy X resulted in increasing level of income of the farmers due to the improved access to a local pricing system, ***but it also resulted in elimination and substitution of the services of local produce brokers resulting in ….*** |

## CONCLUSION

One of the most important issues that researchers and practitioners must address is that of a chosen granularity of a project. In the case of academics, for example, it concerns the level of investigation and, correspondingly, the level of research questions that are going to be stated. For all intents and purposes this deals with selecting along the "general-to-specific" (e.g., grand theory to case study) spectrum of an inquiry. If, let us say, a research question deals with the finding an impact of X on Y and the impact is found to be significant, then it is only so for the level of the context of the study. If an investigation is context-specific, then the findings also are. What was found positive for the wealthy may not be so for the poor (Harris, 2016).

Similarly, practitioners working in the area of ICT deal with projects introducing a change into a particular context, and the context could be very narrow (e.g., level of a department), or it could be very broad (e.g., level of economy). The chosen context carries a corresponding level of scrutiny regarding the success of the project. So, if an implementation of a new software for HR Department was a successful one, then the designation of a "successful project" was based on the stated in advance criteria (e.g., let us say, satisfaction of user requirements) specific to the context of the project. The success of the project at the level of a department may bring negative consequences along the "general-to-specific" spectrum – it may negatively impact an employee of the HR Department, and it could bring some negatives at the level of the firm that houses the department.

There is a general similarity between researchers and practitioners working in the area of ICT, and this similarity is in terms of the impact of ICT. Most of the time, the investigators are searching for a positive impact of ICT, and, all the time practitioners have in mind obtaining a positive impact of ICT. The problem is, of course, is that of context. And the question is, at what level?

Any introduction of change results in an impact. And it is impossible to have a positive impact without some negative aspects manifesting themselves here and there. If we operate in a single context, which is akin to having a tunnel vision, then we may not see the negatives lurking "here and there". Interestingly, despite the fact that the tools allowing for a multi-level consideration of the context were available to researchers and practitioners for quite some time (Jackson, 1982; Arnold & Wade, 2015), we don't see their widespread application.

## REFERENCES

Arnold, R., & Wade, J. (2015). A definition of systems thinking: a systems approach. *Procedia Computer Science*, *44*, 669–678. https://doi.org/10.1016/j.procs.2015.03.050.

Bosamia, M. P. (2013). Positive and negative impacts of information and communication technology in our everyday life. In *Proceedings of International Conference on Disciplinary and Interdisciplinary Approaches to Knowledge Creation in Higher Education: Canada & India (GENESIS 2013).*

Harris, R. (2016). How ICT4D research fails the poor. *Information Technology for Development*, 22(1), 177–192. https://doi.org/10.1080/02681102.2015.1018115.

Jackson, M. (1982). The nature of "soft" system thinking: the work of Churchman, Ackoff, and Checkland. *Journal of Applied Systems Analysis*, *9*, 17–29.

Tarafdar, M., DArcy, J., Turel, O., & Gupta, A. (2015). The dark side of information technology. *MIT Sloan Management Review*, *56*(2), 61–70.

Tarafdar, M., Gupta, A., & Turel, O. (2013). The dark side of information technology use. *Information Systems Journal*, *23*(3), 269–275. https://doi.org/10.1111/isj.12015.

# Section IV

# Appendix X

## THE PURPOSE AND THE SUGGESTED USE OF THE CONTENT IN THIS APPENDIX

The content presented in this section of the book is intended to benefit, primarily, researchers working in the area of Information and Communication Technologies for Development (ICT4D). Based on our experience, there are three preliminary questions that a ICT4D investigator must address in the beginning of the paper in order to elicit positive responses of the reviewers and editors, as well as in order to incorporate her work within a nomological network weaved by previous investigations in that area.

The first question is:

*What is the model/theory of economic development that underlies the inquiry?*

An implicit assumption that we relied on in our work is that economic development is a necessary condition for a broader human and societal development to take place – may be an insufficient condition on its own, granted, but a necessary one at that. The content presented in X1 should help an investigator in selecting an appropriate to the subject and context of her study model of economic development.

The second question is:

*What is the research framework/model of the study?*

Fundamentally, research questions of any investigation should not come out of the blue – instead, a set of research questions is generated based on the selected (or constructed) framework (research model) of the study. Needless to say, in the presence of a theory supporting the investigation, the research framework/model must be demonstrated to be consistent with the selected theory. This way, the investigator is able to show to her readers, reviewers, and editors alike that the research questions of the study were generated based on the research framework/model of the investigation, where the framework is itself consistent with the theory underlying the inquiry. The content presented in X2 may assist an investigator by illustrating the way of selecting or generating a consistent with a theory research framework/model.

# X1 Models of Economic Growth

One of the important requirements for selecting a theory of economic development within the context of this text is that it should allow for considering ICT being one of the factors of economic growth. This requirement allows us to remove from consideration such candidates as the early theories of economic development espoused by Adam Smith (where the specialization via division of labor and exchange are the mechanisms for economic growth) and Karl Marx (where public property and planned economy serve as the engines of development).

One of the key characteristics of ICT is its ability to offer less developed economies an opportunity to develop rapidly, jumping levels and bypassing the usual milestones (e.g., developing cell networks without prior development of landline networks) – this implies that economies may follow their own unique paths to development. Unfortunately, this makes the Linear Stages of Growth' economic models (e.g., Modernization Theory) unsuitable for the purpose of supporting our inquiry, because, based on the fundamental assumption of the model, every economy must pass through the same phases (e.g., the traditional society, the preconditions for take-off, the take-off, the drive to maturity and the age of high mass consumption (Rostow, 1962)), stage by stage.

It has been established that an implementation of ICT requires development of the labor force – it would appear that we could adopt Structural Change models as an underlying framework for our purposes. After all, steady accumulation of physical and human capital, paired with savings and investments, serve as drivers of economic growth for the proponents of the theory. However, Structural Change models deal with the reallocation of labor from the agricultural to the industrial sector as being the driver of economic growth. This has two important implications. First, implementation of ICT is associated not with industrial, but with post-industrial sector; and second, the policies of shifting labor force from agricultural sector in developing countries have been widely recognized as having an overall negative effect (World Bank, 2000). Thus, we remove this model from further consideration.

Implementation of ICT in an economy is often perceived as a collaborative intercountry effort, where, quite often, wealthier counterparts and international agencies help their less developed peers. Perhaps, International Dependence models (Heller, Rueschemeyer, & Snyder, 2009) could be adopted to support our inquiry. However, the main premise of the associated theory is that the less developed countries are taken advantage of by their more developed, wealthier counterparts – consequently, a lack of development is due to the dominance of developed countries over developing countries, and the path to success for developing countries is via the route of ending the economically detrimental relationships with the developed world and by closing their borders and economies to the developed countries. History, however, demonstrated that it is the emphasis on cooperation and trade with the advanced economies (which is contrary to autarkical model espoused by the theory) that resulted in economic success for developing economies (e.g., Hong Kong, Singapore, Taiwan and South Korea, during the 1970s and 1980s) – this resulted in decline of popularity of the theory in 1980s. Unsurprisingly, we consider International Dependence theory as not being suitable for the purposes of our study.

Contrary to the perspectives of the International Dependence model, Neoclassical theories (Hahn, 2010) blame domestic, not international issues for the lack of economic development (au contraire, foreign aid and foreign trade are some of the drivers of the development). Instead of external factors, it is internal issues of poor resource allocation, price distortion and corruption that apply the brakes to economic development (Meier & Stiglitz, 2000). We must note that the above mentioned problems, at least in part, are related to the issue of information transparency, which is successfully dealt with via application of ICT. A free market line of thinking within Neoclassical theories is a Traditional Neoclassical Growth model (framework of Neoclassical Growth Accounting, a.k.a. Solow' model), where economic growth is impacted by quality and quantity of labor, capital investments, and technological progress. However, social, political, and cultural aspects also play an important role (let us recall that this perspective embraces free market), and the economic development could fail to materialize to its fullest extent due to inadequate legal and regulatory framework. Despite its shortcomings, the Neoclassical Growth model fits well with the purposes of our inquiry for the following reasons:

- The focus of the model is internal to the economy – the sources of growth reside under the purview of policy makers of a given economy.
- The constructs of the model could be uniformly represented across a variety of economies.
- The constructs of the model could be objectively reflected by commonly accepted measures.
- The model associates the barriers to economic growth with limitations of information channels and information transparency – aspect easily impacted by application of ICT.
- The model considers the importance of an adequate governmental services to economic development – and ICT has been shown to be a useful tool in improving quality of governmental services.

While Solow's growth model considers one of the sources of growth – technological progress – to be an exogenous (external) factor, the New Growth theory (Islam, 2004) explicitly links, within the model, technological progress to the production of knowledge as a driver of sustained economic growth. Within this framework, the process of knowledge creation cannot be left to individuals, because the individual would not be able to capture all the gains associated with new knowledge creation. Thus, the role of government and public policies is emphasized as a tool in formation of a human capital and encouragement of foreign private investments in knowledge-intensive industries, such as those associated with ICT. The New Growth theory would seem a perfect candidate to support our inquiry, but two points of concern stop us from adopting it as a foundation of the study. First, developing economies may not be a good target for a model that emphasizes a knowledge-based growth (Kaur & Singh, 2016), for it has been noted that in less developed economies there are many other factors that are more likely to drive the economic development (Cornwall & Cornwall, 1994; Parash, 2015). Second, any meaningful application of the theory requires representation and capturing of the concept "knowledge creation", and, given the context of the study, it would be an unsurmountable challenge to overcome.

Unlike previously covered models, the theory of Coordination Failure (Hoff, 2000; Kydd & Dorward, 2004) does not have a fixed culprit responsible for the lack of economic development. It is not Adam Smith's division of labor and market exchange, it is not public property and planned economy of Marx, it is not "must go through" sequence of Linear Stages, it is not a better developed oppressor of International Dependence, it is not reallocation of labor of Structural Change, and it is not absence of capital investments, labor, and technical progress of Neoclassical Growth model, neither is it the knowledge creation of the New Growth theory. Instead, it is a market's failure to coordinate complementary activities that is responsible for the lack of economic development – simply put, in the absence of coordination, market settles on a sub-optimal result. A possible cure to such situation is the emphasized role of government (because if left on their own, markets settle on local maxima versus converging on a global maximum), and of its supervision over public-driven massive investment program geared toward creating complementarities in the market. From one perspective, Coordination Failure theory is well-suited for the purpose of investigating "anything ICT" – after all, ICT is introduced as a complementarity (e.g., IT complements the skills of an employee) and is intended to produce complementarities along the way (e.g., IT creates and introduces new jobs in newly created sectors – cybersecurity, data analytics, etc.). Yet, from another perspective, this theory is almost impossible to apply for our purposes, for the following reasons:

- The scope of ICT-related complementarities within an economy is hard to define, to uniformly describe, and impossible to measure on a common scale.
- The government-centric "central planning" approach to managing complementarities is hardly applicable to the context of the study.
- The "big push" model of government-supervised investments may not be supported but what is actually happening in developing economies. For one, we do not, and did not, see a large scale "big push" investments in ICT.

There is no single agreed-upon theory of development – each model considers a few relevant dimensions without claiming exclusivity of knowledge or a gnostic insight on the topic. Behind the diversity of perspective on the issue of economic growth there are three common questions that every model tries to deal with. The first question is how to explain economic development over time. The second question is how to identify barriers to growth. And the third question is what is the appropriate role of a government in the process.

In order to support our line of inquiry, we identified the framework of Neoclassical Growth as a suitable candidate – we hope that, given the reviewed above alternatives, our readers would concur with the selection made. In simple terms, the framework of Neoclassical Growth Accounting explains the economic growth in terms of capital, labor, and technological progress, where governmental policies of liberalization, stabilization and privatization are central elements of the development effort.

We must note that the selected theoretical support is not exclusive to other perspectives on development. For example, the Neoclassical Growth model is perfectly compatible with and complementary to the Development as Freedom' perspective of Amartya Sen (1999). Let us consider two aspects of the selected model – investments in ICT as a factor of economic development and macroeconomic growth. ICT Capabilities created as a result of investments in ICT are perfectly available, via creating new and optimizing existing information channels, of addressing some of the issues associated with political freedoms and transparency of the relationships between people and people, people and government, and people and social and political groups. Macroeconomic growth, on the other hand, is capable of addressing a freedom of opportunity via generating additional wealth and affording a greater access for poor to financing and credit. Similarly, additional wealth generated as a result of economic development may offer a greater level of economic protection for the poor via more money being available to allocate to income supplements and unemployment benefits. It is perfectly reasonable, consequently, to see Neoclassical Growth model as describing how to generate wealth, and to see Sen's model as describing how to allocate generated wealth in order to address important problems of social and human development.

## REFERENCES

Cornwall, J. and Cornwall, W. (1994). Growth Theory and Economic Structure. Economica, 61(242), 237–251.

Hahn, F.H. (2010). Neoclassical Growth Theory. In: Durlauf S.N., Blume L.E. (eds.) Economic Growth. The New Palgrave Economics Collection. Palgrave Macmillan, London.

Heller, P., Rueschemeyer, D. and Snyder, R. (2009). Dependency and Development in a Globalized World: Looking Back and Forward. Studies in Comparative International Development (SCID), 44(4), 287–295.

Hoff, K. (2000). Beyond Rosenstein-Rodan: The Modern Theory of Underdevelopment Traps. World Bank Development Economics Conference 2000.

Islam, N. (2004). New Growth Theories: What Is in There for Developing Countries? The Journal of Developing Areas, 38(1), 171–212.

Kaur, M. and Singh, L. (2016). Knowledge in the Economic Growth of Developing Economies. African Journal of Science, Technology, Innovation and Development, 8(2), 205–212.

Kydd, J. and Dorward, A. (2004). Implications of Market and Coordination Failures for Rural Development in Least Developed Countries. Journal of International Development, 16, 951–970.

Meier, G. and Stiglitz, J. (2000). Frontiers of Development Economics: The Future in Perspective. New York: World Bank and Oxford University Press.

Parash, U. (2015). Factors Affecting Economic Growth in Developing Countries. Major Themes in Economics, 17, 37–54.

Rostow, W. W. (1962). The Process of Economic Growth. Clarendon Press, Oxford.

Sen, A. (1999). Development as Freedom (1st ed.). Oxford University Press, New York, NY.

World Bank. (2000). Entering the 21st Century – World development report 1999/2000. Oxford University Press, New York, NY.

# X2 A Model of the Socio-Economic Impact of ICT

An overview provided in the previous part of the text, capped by the selection of the preferred theoretical framework, allows us to formulate a conceptual model according to which ICT contributes to economic development – it is as follows:

> *(ICT Labor, ICT Capital/Investments, ICT-driven Technological Progress) → Macroeconomic Growth.*

A reliance on such grand theory as *Neoclassical Growth* model allows for linking the factors of economic growth to the growth itself, but it does not shed any light on the *mechanisms* by which, let us say, labor and investments work their ways to contributing to macroeconomic growth.

As preamble to the discussion of the link between ICT and macroeconomic growth we need to establish a connection between macroeconomic development and factors of growth – investments and labor. The obvious connection would be that a higher level of economic development allows for higher levels of investments and higher quality of the labor force – this, over time, results in the accumulation of labor – as well as a build-up of capital-related ICT infrastructure. But, on their own, accumulated capital, investments, and available labor are not very useful – they are simply resources waiting to be taken advantage of. Simply put, increasing levels of investments in ICT fueled by economic development, *applied on top of the existing socio-technical ICT infrastructure*, create increasing levels of *ICT Capabilities*.

The *Networked Readiness Index* (NRI) serves as an established measure of the *ICT Capabilities*, or, *drivers*, of socio-economic development (World Economic Forum, 2016). Since its development in 2002 the NRI framework has been highly regarded as an authoritative source of data and an assessment tool geared towards global leaders, practitioners, policy makers (Kirkman, Osorio, & Sachs, 2002), which was also available to academic researchers (Indjikian & Siegel, 2005; Ifinedo, 2005; Vehovar et al., 2006; Wielicki & Arendt, 2010; Chanyagorn & Kungwannarongkun, 2011; Milenkovic et al., 2016; Samoilenko & Osei-Bryson, 2017, 2019).

The NRI framework used in this text is reflected by a single composite index comprised of four sub-indexes – three *drivers* (*Environment*, *Readiness*, and *Usage* sub-indexes) and one *Impact* sub-index. The sub-indexes are represented by 10 sub-categories. The values of the index, sub-indexes, and sub-categories are represented by scores ranging from 1, being the lowest, to 7, being the highest – this allows not only for a cross-country comparison, but also for comparison of each country vis-à-vis itself over time.

One of the underlying principles of the framework is that "NRI should provide clear policy guidance" (World Economic Forum, 2016, p. xi) to its users, which is not an easy undertaking to implement because within the framework "…the complex relationships between ICTs and socio-economic performance are not fully understood and their causality not fully established" (Di Battista, Dutta, Geiger, & Lanvin, 2015, p.4). Unsurprisingly, the framework of NRI conceptualizes and captures *ICT Capabilities* in terms of three *Drivers – Environment, Readiness*, and *Usage*. It is worth noting again – ICT capabilities are created on top of the existing socio-technical infrastructure. This allows us, so far, to establish and justify the following chain of links:

*Economic Development → Investments in ICT (applied to existing socio-technical ICT infrastructure) → ICT Capabilities → Economic Development.*

We must note that while it is inevitable that investments in ICT would work via creation of *ICT Capabilities*, investigators have options in representing the *ICT Capabilities* construct (Zuppo, 2012; Kleine, 2013). In this chapter, we selected the framework of NRI for the purpose because it offers an established representation that is widely accepted by scholars and practitioners, and also could be applied to our context – two different sets of economies. However, in a narrower context investigators have been known to "unpack" the meaning and representation of the construct to elucidate transformation paths by which investments make their ways to becoming micro- and macroeconomic benefits. For example, Samoilenko (2016) traced the *investments in ICT to economic development* path via a chain of links starting from the introduction of Telecom products, to decreasing cost of acquisition and utilization of Telecom products, and then to the resultant increase in the level of consumption.

Another advantage afforded by a reliance on the NRI framework is that it includes an *Impact* sub-index, which represents and measures social and economic impacts of the *Drivers – ICT Capabilitie*s created as a result of investments in ICT. However, within the framework "*…the complex relationships between ICTs and socio-economic performance are not fully understood and their causality not fully established*" (Di Battista, Dutta, Geiger, & Lanvin, 2015, p.4). Consequently, *Drivers* of NRI are not causally linked to *Impact* – this poses a question: *How do Drivers produce Impact?*

An answer was offered by Samoilenko (2018), who suggested that *ICT Capabilities* (e.g., *Drivers*) must enable some sort of a socio-economic *engine* that actually produces *Impact*, and the consequent framework was developed by Samoilenko and Osei-Bryson (2019). Importantly, this model of "*Drivers → Engine → Impact*" was then reconciled with the *Neoclassical Growth* model using a framework of sustainable macroeconomic impact of investments in ICT (Samoilenko, 2016) – the resultant model is depicted by Figure X2.1.

The model depicted in Figure X2.1 supports the *development cycle* of the study, which makes the theoretical foundation of our investigations to be consistent not

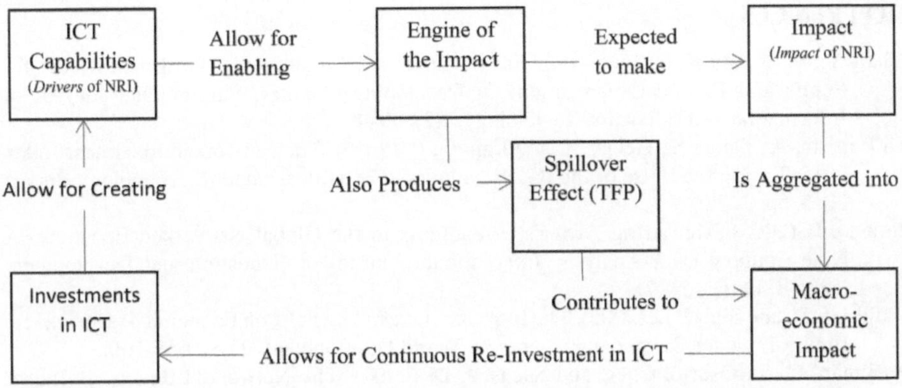


**FIGURE X2.1** Framework of NRI and *Neoclassical Growth* model (Samoilenko & Osei-Bryson, 2021).

only with the established NRI framework, but with the grand theory of *Neoclassical Growth* as well, without making it inconsistent with other perspectives on development (e.g., Sen, 1999).

In 2020 the NRI released a new model that, while maintaining a continuity with the previous version, is better geared towards reflecting current and future ICT-related issues. The framework still consists of three *drivers* and one impact, but the names and representations of sub-indexes change – the new model is comprised of four pillars: *Technology*, *People*, *Governance*, and *Impact*. The changes in names or representations do not impact the overall structure of the *Drivers → Engine → Impact* model of Samoilenko and Osei Bryson (2019) – we present their framework with the updated version of NRI in Figure X2.2.

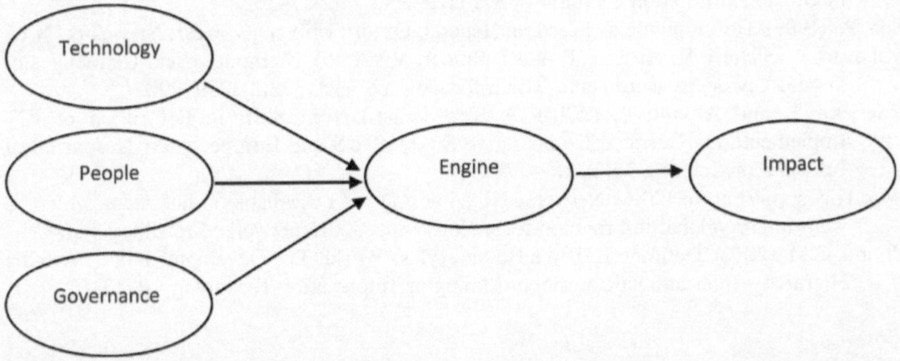


**FIGURE X2.2** *Drivers → Engine → Impact* model using the 2020' version of NRI.

## REFERENCES

Chanyagorn, P. and Kungwannarongkun, B. (2011). ICT Readiness Assessment Model for Public and Private Organizations in Developing Country. International Journal of Information and Education Technology, 1(2), 99–106.

Di Battista, A., Dutta, S., Geiger, T. and Lanvin, B. (2015). The Networked Readiness Index 2015: Taking the Pulse of the ICT Revolution, Global Information Technology Report 2015, pp. 3–28.

Ifinedo, P. (2005). Measuring Africa's E-readiness in the Global Networked Economy: A Nine-Country Data Analysis. International Journal of Education and Development using ICT, 1(1), 53–71.

Indjikian, R. and Siegel, D. (2005). The Impact of Investment in IT on Economic Performance: Implications for Developing Countries. World Development, 33(5), 681–700,

Kirkman, G. S., Osorio, C. A. and Sachs, J. D. (2002). The Networked Readiness Index: Measuring the Preparedness of Nations for the Networked World. The Global Information Technology Report: Readiness for the Networked World, 10-30. Oxford University Press, New York, NY.

Kleine, D. (2013). Technologies of Choice? ICTs, Development, and the Capabilities Approach. The MIT Press, Cambridge, MA, London, England.

Milenkovic, M. J., Brajovic, B., Milenkovic, D., Vukmirovic, D. and Jeremic, V. (2016). Beyond the Equal-weight Framework of the Networked Readiness Index: A Multilevel I-Distance Methodology. Information Development, 32(4), 1120–1136.

Samoilenko, S. (2016). Where Do Investments in Telecoms Come from? Developing and Testing a Framework of Sustained Economic Impact of Investments in ICT. Journal of Information Technology for Development, 22(4), 584-605.

Samoilenko, S. (2018). Socio-Economic Impact of ICT-Enabled Public Value in Sub-Saharan Economies. In Proceedings of 6th Mediterranean Interdisciplinary Forum on Social Sciences and Humanities, MIFS 2018, 24–25 May 2018, Barcelona, Spain.

Samoilenko, S. and Osei-Bryson, K.M. (2017). An Analytical Framework for Exploring Context-Specific Micro-Economic Impacts of ICT Capabilities. Journal of Information Technology for Development, 24(2), 633–657.

Samoilenko, S. and Osei-Bryson, K.M. (2019). Representation Matters: An Exploration of the Socio-Economic Impacts of ICT-Enabled Public Value in the Context of Sub-Saharan Economies. International Journal of Information Management, 49, 69–85.

Samoilenko, S. and Osei-Bryson, K.M. (2021). ICT Capabilities and the Cost of Starting Businesses in Sub-Saharan African Economies: A Data Analytic Exploration, Journal of Global Information Technology Management, 24(1), 7–36.

Sen, A. (1999). Development as Freedom (1st ed.). Oxford University Press, New York, NY.

Vehovar, V., Sicherl, P., Hüsing, T. and Dolnicar, V. (2006). Methodological Challenges of Digital Divide Measurements. The Information Society, 22(5), 279–290.

Wielicki, T. and Arendt, L. (2010). A Knowledge-Driven Shift in Perception of ICT Implementation Barriers: Comparative Study of US and European SMEs. Journal of Information Science, 36(2), 162–174.

World Economic Forum. (2016). Networked Readiness Dataset. Available on-line at: http://reports.weforum.org/global-information-technology-report-2016/networked-readiness-index/

Zuppo, C.M. (2012). Defining ICT in a Boundaryless World: The Development of a Working Hierarchy. International Journal of Managing Information Technology, 4, 13–22.

# Index

*Italicized* pages refer to figures and **bold** pages refer to tables.

## D

## N



## O



## P



## R